1 Introduction

Nanomaterials are substances engineered at the atomic or molecular scale, typically less than 100 nm in size [1]. Compared to their bulk counterparts, nanomaterials exhibit distinct physical and chemical properties, including an increased surface area [2], enhanced reactivity, durability, permeability [3], altered optical properties and improved mechanical strength, making them highly versatile and valuable across a wide array of applications [2]. The potential of nanomaterials extends beyond conventional boundaries, enabling ground-breaking solutions in health and cosmetics [3], as well as environmental sustainability and technology [4].

The cornerstone of these advancements in various fields lies in the meticulous synthesis and precise processing of nanomaterials, which govern their functionality and application potential. The synthesis methods of nanomaterials fall into two categories: top-down and bottom-up. Top-down techniques, such as ball milling [5] and lithography [6], break down bulk materials into nanoscale pieces [7]. Bottom-up approaches, such as chemical vapour deposition and sol–gel processes, build nanomaterials atom by atom, offering better control over size, shape and composition [8]. Selecting the appropriate processing method is crucial, as it significantly impacts the properties and applications of nanomaterials. This chapter explores the diverse methods of synthesising and processing nanomaterials, as well as the properties of nanomaterials utilised in various fields and applications.

2 Processing of Nanomaterials

Understanding the preparation and processing of nanomaterials is pivotal for harnessing their full potential across various applications. Advanced synthesis techniques, broadly categorised into top-down and bottom-up approaches, have been developed to precisely control nanoparticle size, shape and composition [9]. Top-down methods focus on breaking down bulk materials into nanoscale particles, and bottom-up approaches construct nanomaterials atom by atom or molecule by molecule. Additionally, green synthesis methods have gained prominence as eco-friendly alternatives, utilising biological and natural resources to sustainably produce nanomaterials. These methodologies lay the foundation for the diverse functionalities and transformative applications of nanomaterials.

2.1 Top-Down Approaches

Top-down approaches are methods that start with bulk materials and break them down into smaller nanoscale particles. These techniques, including ball milling, lithography, sputtering, laser ablation and arc discharge, rely on physical or mechanical processes to reduce material size. This approach offers precise control over particle size and shape by applying external forces such as grinding, irradiation, or etching. Top-down approaches are highly efficient in producing nanomaterials from larger precursor materials, making them particularly suitable for large-scale production.

Ball Milling

Ball milling is a fundamental technique in nanomaterial processing, renowned for its ability to modify the properties and enhance the functionalities of nanomaterials [10]. This technique employs a rotating cylindrical vessel filled with balls made of hard materials such as steel or ceramics (Fig. 1a). As the vessel rotates, the balls collide with the bulk material, causing it to break down. The rotations generate intense mechanical forces that induce repeated deformation, rewelding and fragmentation of the material, resulting in finely dispersed nanomaterials embedded within a grain-refined structure (Fig. 1b) [11, 12]. The process parameters, including the speed of rotation, ball size and milling duration, can be meticulously controlled to achieve the desired particle characteristics [13].

Two mechanisms influence the characteristics of the material during the milling process: cold welding and fragmentation. Cold welding (the first mechanism) increases the average size of composite particles, while fragmentation (the second mechanism) breaks them into smaller components [5]. Over time, these opposing processes reach a steady-state equilibrium, where a balance is achieved, resulting in a uniform distribution of nanomaterials with optimised properties [14]. This dynamic balance highlights the effectiveness of ball milling as a versatile technique for nanoparticle synthesis. Besides the ability to produce nanoscale materials independently, ball milling is frequently used as a preparatory step before extraction and other processes. Ball milling effectively reduces material size and improves

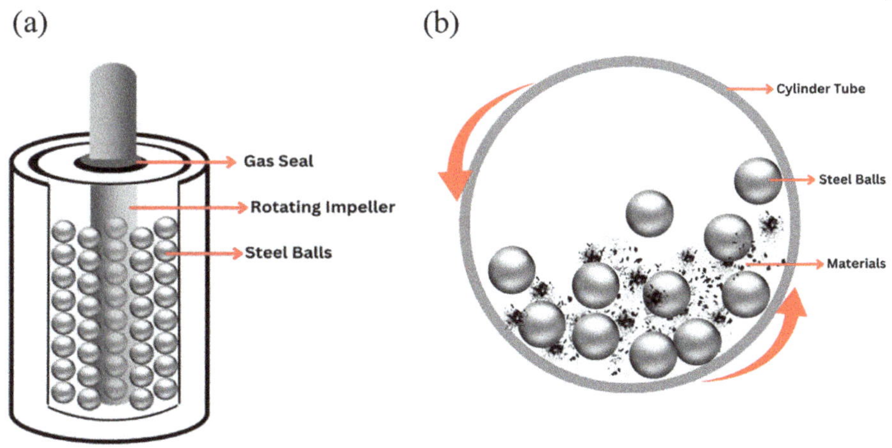

Fig. 1 Ball milling **a** rotating cylindrical vessel and **b** cross-section of the vessels

processing outcomes [15]. The intense grinding at high speeds enhances the diffusion of bioactive compounds and promotes cell wall breakdown, thereby facilitating further processing steps [11]. However, the nanomaterials produced from this process usually have a broader size range (3–25 nm), which limits their functionality and applications when a more precise size control is required [16].

Lithography

Lithography is another technique in nanomaterials processing that creates precise patterns on the substrate, which is essential for producing nanomaterials. Unlike ball milling, lithography achieves tighter size distributions, making it more reliable for size-controlled applications. This method involves coating the substrate with a light-sensitive material called a resist, enabling the formation of intricate nanostructures (Fig. 2). A pattern is then transferred onto the resist using various methods such as light (photolithography) [17], electron beams (electron beam lithography) [18], or moulds (nanoimprint lithography) [19]. After exposure, the resist is developed, revealing the pattern on the substrate. This pattern can then be used to etch or deposit materials, thus forming intricate nanostructures. Lithography is crucial in fields like electronics, photonics and materials science, enabling the creation of devices with nanoscale features [6, 19]. Examples of nanomaterials produced using these techniques include carbon nanotubes (CNTs), cylindrical tubes made of rolled sheets of carbon atoms, known for their exceptional strength and electrical conductivity [20].

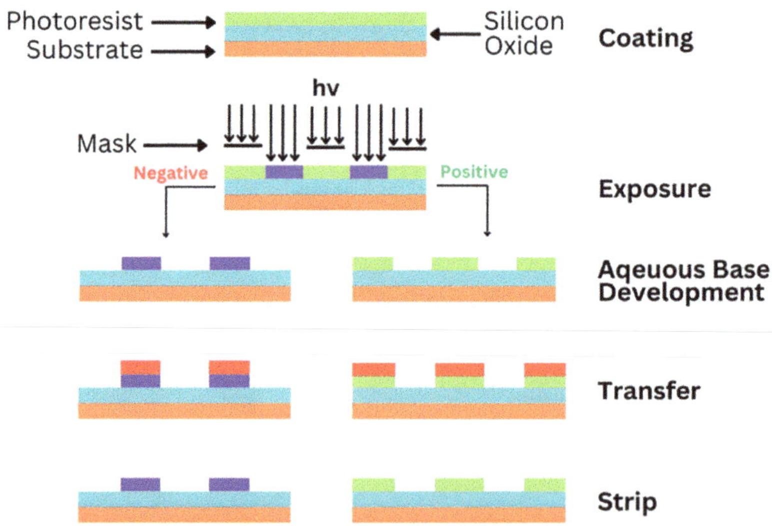

Fig. 2 Nano-pattern transfer on a substrate by photolithography

Sputtering and Laser Ablation

Sputtering and laser ablation are other commonly used techniques in nanomaterial processing. Sputtering involves ejecting atoms from a target material using high-energy particles, which are then deposited onto a substrate to form thin films [21]. In contrast, laser ablation uses a high-energy laser beam to vaporise the target material, resulting in the creation of nanoparticles [22]. Both techniques enable precise control over the properties of the resulting nanomaterials, making them essential for creating various nanostructures for advanced applications [23].

The sputtering technique begins with placing both the substrate and the target material inside a vacuum chamber filled with an inert gas, such as argon. A high voltage is then applied, creating a plasma of ionised gas atoms. These ions are accelerated towards the target material, colliding with its surface and transferring momentum (Fig. 3a). This impact causes atoms from the target material to be ejected in a process known as sputtering [23, 24]. The ejected atoms then travel through the vacuum and deposit onto the substrate, forming a thin film with controlled thickness and composition by energetic ions, typically in a vacuum chamber [24].

One of the key advantages of sputtering is its ability to provide precise control over film thickness, composition and morphology [25]. By adjusting parameters such as gas pressure, power input and substrate positioning, researchers can tailor the film's properties to meet specific requirements. Additionally, sputtering allows for the deposition of multi-layered structures, co-sputtering of different materials, and doping to modify material characteristics [26]. Despite its advantages, the technique produced a wider size range of nanomaterials (1–100 nm) and requires careful optimisation of conditions to ensure uniformity, avoid contamination and achieve the desired film properties [27].

Laser ablation also utilises the concept of sputtering materials (Fig. 3b). The technique is usually used to remove material from a solid target through irradiation

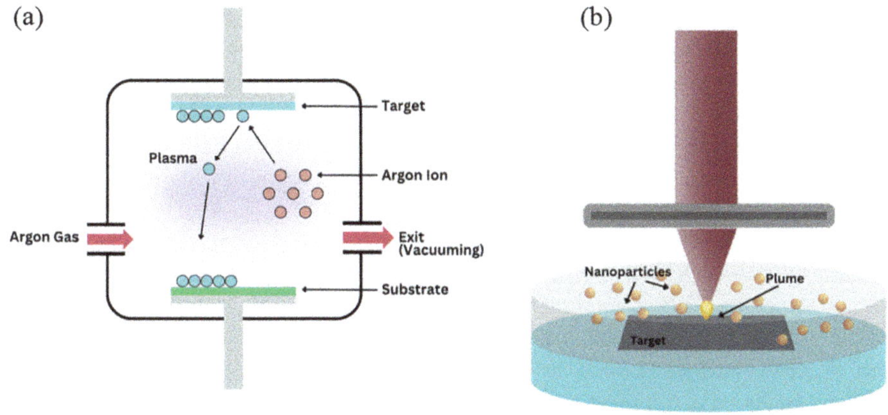

Fig. 3 Nanomaterial processing via **a** the sputtering process to produce nano-film and **b** the sputtering of nanomaterials from laser ablation

with a high-powered laser beam. This process begins when a pulsed laser, such as neodymium-doped yttrium aluminium garnet (Nd:YAG) or excimer laser, is directed at the target material [28]. The high energy of the laser pulses rapidly heats and vapourises the surface of the target, initiating material removal. As a result, a plasma plume consisting of ionised atoms and molecules from the target material is generated [29]. The expansion and cooling of this plasma lead to nucleation, where vapourised species aggregate to form nanoparticles [30]. Depending on the specific setup and intended application, these nanoparticles can be collected on a substrate or suspended in a liquid medium.

The effectiveness of laser ablation depends on various factors, including laser wavelength, pulse duration, energy density and ambient conditions (i.e. the presence of a buffer gas or liquid medium) [31]. Shorter laser pulses and higher energy densities typically enhance nanoparticle production, while the surrounding environment influences the synthesised nanomaterial's composition, size and morphology [22]. Additionally, by adjusting experimental parameters, researchers can tailor the properties of the resulting nanoparticles, ensuring control over their size distribution and purity. One of the significant advantages of laser ablation is its ability to synthesise high-purity nanoparticles without requiring chemical precursors, making it an attractive method for applications requiring minimal contamination [32]. Although laser ablation can produce comparatively smaller nanomaterial sizes (< 10 nm), precise control of laser parameters and target properties is essential to optimise nanoparticle yield and consistency. Improper settings can result in irregular particle sizes or incomplete ablation [31].

Arc Discharge Method

The arc discharge method is commonly employed for synthesising carbon-based nanomaterials, particularly carbon nanotubes (CNTs) and fullerenes [33]. This approach involves positioning two carbon electrodes close to an inert gas atmosphere (i.e. argon or helium) to prevent unwanted reactions (Fig. 4). A metal catalyst is often introduced to facilitate the selective formation of specific nanostructures. When a high current is applied, an arc discharge occurs between the electrodes, generating extreme heat that vaporises the carbon material [34]. As the vaporised carbon cools and condenses, it forms various nanostructures, including single-walled carbon nanotubes (SWCNTs) [35], multi-walled carbon nanotubes (MWCNTs) [36] and fullerenes. These nanomaterials are then collected from the chamber walls and electrode surfaces.

Although arc discharge typically produces nanomaterials in the range of 1 to 100 nm, it is capable of yielding high-quality, crystalline nanomaterials with well-defined structures [37]. Compared to other synthesis techniques, it offers a relatively simple setup while yielding high-purity nanostructures [29]. However, precise control over process parameters such as gas pressure, electrode composition, and current intensity is crucial to optimise product uniformity and minimise unwanted byproducts like amorphous carbon [38].

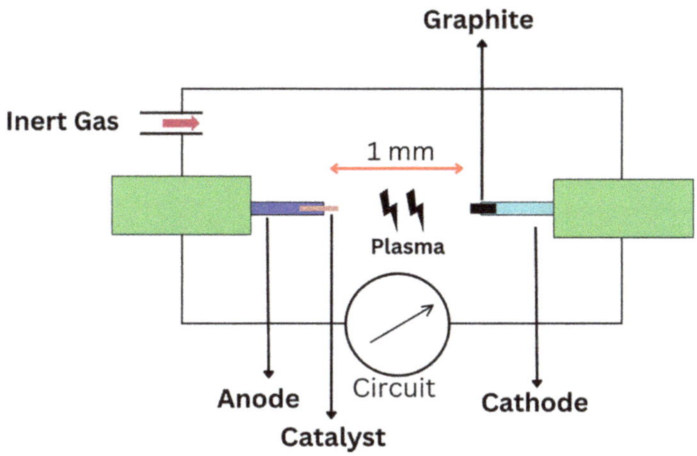

Fig. 4 Diagram of arc discharge nanomaterial processing

2.2 *Bottom-Up Approaches*

Nanotechnology has opened up new possibilities by allowing scientists to manipulate materials at the atomic and molecular levels. One of the key strategies in this field is the bottom-up approach, which involves building nanostructures from the smallest units, the atom and molecules. This method contrasts with the top-down approach, which starts with bulk materials and reduces them to nanoscale dimensions. Bottom-up approaches, such as chemical vapour deposition (CVD), hydrothermal and solvothermal methods, sol–gel synthesis and co-precipitation, offer precise control over the properties of nanomaterials. This technique is pivotal in driving innovations across various fields, including electronics, photonics and medicine. Researchers can create advanced nanomaterials with unique functionalities and enhanced performance by assembling materials from the ground up.

Chemical Vapour Deposition (CVD)

One of the most common bottom-up methods in processing nanomaterials is chemical vapour deposition (CVD). CVD involves exposing a substrate (the surface on which the materials are deposited) to one or more volatile precursors. These precursors react and/or decompose on the substrate's surface to produce the desired deposit (Fig. 5). The byproducts of these reactions are usually gases that are removed from the reaction chamber [39]. CVD encompasses various techniques designed to optimise specific applications and performance characteristics, such as atmospheric pressure CVD (APCVD), low-pressure CVD (LPCVD), plasma-enhanced CVD (PECVD) [40] and aerosol-assisted CVD (AACVD) [41], which provide distinct advantages, including the versatility of the nanomaterials produced.

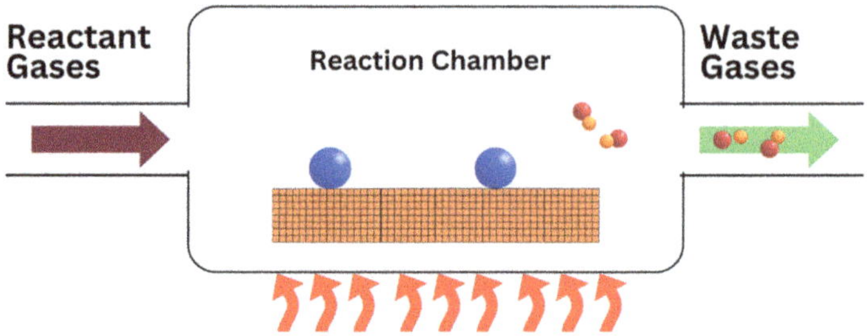

Fig. 5 Typical pathways of gases during the CVD process

CVD presents numerous advantages, including the fabrication of high-purity materials, the ability to achieve uniform coatings on intricate surfaces, and scalability for large-scale production [40]. However, implementing these techniques poses several challenges. High processing temperatures can restrict substrate selection, and the requirement for sophisticated equipment and stringent process control adds to the complexity [8, 39]. Additionally, the nanomaterials produced often exhibit a larger size range, which limits their applicability in various fields. Despite these limitations, CVD remains an indispensable technique in advancing nanotechnology and materials science.

Hydrothermal and Solvothermal Method

The hydrothermal synthesis of nanoparticles involves dissolving reactants in water within a sealed autoclave [41]. Unlike hydrothermal, the solvothermal method provides versatility in synthesising a wide range of nanomaterial precursors, as most are soluble in organic solvents. Both methods (Fig. 6a and b) are conducted at elevated temperatures and pressures, which enhances the solubility of the reactants and accelerates the reaction kinetics [42]. The reactions are typically conducted in a sealed vessel capable of withstanding high reaction conditions with temperatures ranging from 100 to 300 °C or exceeding the solvent's melting point and pressures reaching several hundred atmospheres [42, 43]. Under these conditions, the precursors undergo nucleation and subsequent growth, leading to the formation of the target materials. Upon completion of the reaction, the autoclave is gradually cooled, and the synthesised materials are recovered through filtration or centrifugation.

Comparatively, hydrothermal is often preferred for achieving smaller material sizes, as solvothermal methods tend to produce slightly larger nanomaterials (> 200 nm). Additionally, hydrothermal techniques provide high-purity nanoparticles with controlled morphology and excellent crystallinity [44]. Solvothermal methods, while producing larger nanomaterials, offer several advantages, including the production of well-crystallised materials with a controlled morphology. This makes solvothermal methods highly versatile and effective for producing nanomaterials for advanced applications [45, 46].

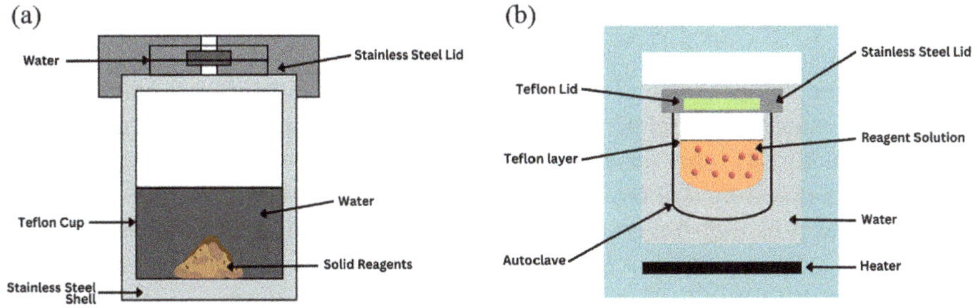

Fig. 6 Diagram of **a** hydrothermal and **b** solvothermal processing of nanomaterials

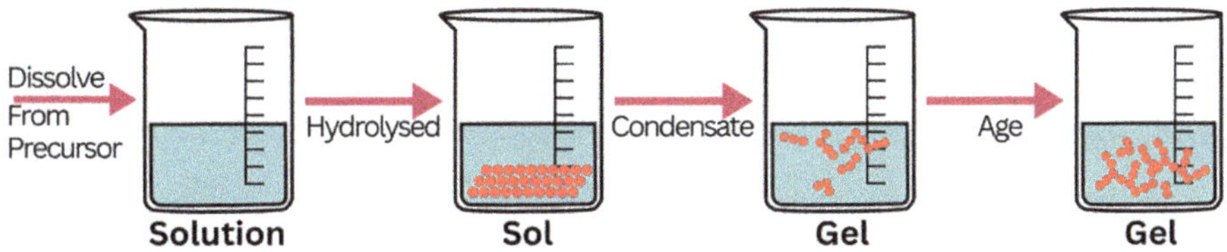

Fig. 7 Sol–gel process of nanomaterials

Sol–Gel Synthesis

Another processing method in the production of nanomaterials is the sol–gel method. Unlike other bottom-up techniques, the sol–gel process entails the transition of a system from a liquid 'sol', characterised as a colloidal suspension of particles, to a solid 'gel' phase. This transformation is governed by hydrolysis and polycondensation reactions [26]. Typically, metal alkoxides serve as precursors, undergoing hydrolysis to generate metal hydroxides. These hydroxides subsequently undergo condensation, forming a three-dimensional network of metal–oxygen–metal (M–O–M) bonds, ultimately forming a gel (Fig. 7).

The sol–gel process initiates with the hydrolysis of metal alkoxides, wherein these precursors react with water to generate metal hydroxides. This initial step is critical, as it establishes the foundation for subsequent reactions (Eq. 1) [26]. Following hydrolysis, condensation reactions occur, facilitating the formation of metal–oxygen–metal (M–O–M) bonds. These bonds contribute to the development of a 3D network, driving the transition from a liquid 'sol' to a solid 'gel' phase. The resulting gel can undergo further processing, including drying and heat treatment, to yield the final solid material. Precise control of reaction conditions throughout the process is crucial for achieving the desired material properties (Eq. 2).

$$\mathrm{M(OR)}_n + \mathrm{H_2O} \rightarrow \mathrm{M(OH)}_n + \mathrm{ROH} \tag{1}$$

$$\mathrm{M(OH)}_n \rightarrow \mathrm{M - O - M} + \mathrm{H_2O} \tag{2}$$

where M = metals; R = organic group (alkyl/aryl) and n = number of (-OH) and (-OR) groups

In short, the sol–gel method offers several significant advantages, including the synthesis of materials at relatively low temperatures, which leads to high purity and homogeneity [47]. Its versatility enables the fabrication of a diverse range of materials with precisely tailored properties. However, the process also presents challenges, such as potential shrinkage and cracking during drying, as well as the necessity for stringent control over reaction conditions to achieve the desired material characteristics [26]. Regarding particle size, the sol–gel method typically produces nanomaterials ranging from 1 to 100 nm, making it suitable for applications that do not require precise material sizes. Despite these limitations, the sol–gel method remains a powerful and widely utilised technique in materials science. It facilitates the development of advanced materials with unique structural and functional properties [48].

Co-Precipitation

Among all the bottom-up methods for producing nanomaterials, co-precipitation is notable for producing a tighter size range (2–50 nm), making it more suitable for size-controlled applications. This method involves dissolving metal salts or other precursors in a solution, followed by the addition of a precipitating agent, such as ammonia or sodium hydroxide [49]. This induces the precipitation of metal ions as hydroxides or other insoluble compounds. The resulting precipitate consists of a uniform mixture of the desired components, which can undergo further processing to obtain the final material with tailored properties [50].

This method offers several advantages. It is a relatively simple and cost-effective technique often conducted at room temperature, making it highly accessible [51]. This method also ensures a homogeneous distribution of components in the final material, which is crucial for applications that require uniformity at the molecular level. Despite its advantages, the co-precipitation method also presents certain challenges. Controlling particle size and morphology with high precision can be difficult, potentially affecting the material's performance [52]. Furthermore, unwanted impurities may co-precipitate alongside the desired material, necessitating additional purification steps. The process parameters, including pH, temperature and mixing rate, must be carefully regulated to achieve reproducible and consistent results [53].

2.3 Green Synthesis

Green synthesis of nanoparticles is an emerging and sustainable approach that harnesses biological systems for nanoparticle production. This method has gained significant attention as an eco-friendly alternative to conventional top-down and bottom-up approaches, which often involve hazardous reagents, high-energy consumption and toxic byproducts [54]. By utilising natural biomolecules from plants, microorganisms, fungi, algae and actinomycetes, green synthesis facilitates the reduction [55] and stabilisation of metal ions in a more sustainable and biocompatible manner [56]. Green synthesis operates under milder conditions, reducing environmental impact and ensuring greater safety for both researchers and industrial processes [54]. The bioactive compounds present in plant extracts [57], microbial metabolites [58] and enzymatic systems act as reducing agents, converting metal precursors into nanoparticles while simultaneously capping and stabilising them to prevent aggregation [56]. This dual functionality enhances nanoparticle stability and improves its use in environmental and medicinal applications.

Plant-mediated green synthesis utilises plant extracts with high concentrations of bioactive molecules as natural reducing and stabilising agents. Various plant parts, including leaves, fruits, roots, stems and seeds, can be used, with extracts prepared by washing, drying and heating plant materials in a solvent (typically water) to release bioactive compounds [59]. These extracts are then mixed with metal salt solutions (e.g. silver nitrate for silver nanoparticles), where phytochemicals facilitate the reduction of metal ions and stabilise the resulting nanoparticles [60]. The process is further optimised by adjusting pH, temperature and reaction time to control nanoparticle size and morphology. Although this method is cost-effective, scalable and provides high-purity nanoparticles with controlled morphology, it is also limited to the variability of plant extract compositions.

Unlike plant-mediated synthesis, which utilises plants' bioactive compounds, microorganism-mediated synthesis utilises metabolic processes from the microorganism and acts as reducing and stabilising agents by secreting enzymes, peptides and polysaccharides. Bacteria, such as *Pseudomonas aeruginosa*, are widely utilised due to their rapid growth and genetic adaptability, which enables the efficient synthesis of silver nanoparticles with antimicrobial and anticancer properties [61, 62]. Fungi like *Aspergillus niger* produce nanoparticles through protein secretion and high metal tolerance, making them effective biocatalysts [63]. These biologically synthesised nanoparticles have diverse applications in medicine, agriculture and environmental remediation. However, challenges such as controlling nanoparticle size, shape and consistency remain, necessitating further research and optimisation to enhance reproducibility and broaden their potential applications [64].

On the other hand, algae-mediated synthesis utilises the photosynthetic properties of algae to produce nanomaterials. Algae utilise sunlight and carbon dioxide to generate energy, making them promising candidates for nanoparticle synthesis. They can function as live-cell factories or provide extracts rich in bioactive compounds that serve as natural reducing and stabilising agents for metal ions [65]. Both microalgae and macroalgae play a role in this synthesis process. Microalgae such as *Chlorella vulgaris* have been used to produce gold nanoparticles due to their high metal uptake capacity [66], while macroalgae like *Sargassum wightii* [67] and *Kappaphycus alvarezii* [52] have demonstrated the ability to synthesise silver nanoparticles. The presence of proteins, polysaccharides and lipids in algae aids in reducing metal ions and stabilising the nanoparticles, ensuring high purity and biocompatibility. Algae-derived nanoparticles have diverse applications in medicine, agriculture and environmental remediation. However, challenges such as maintaining consistent nanoparticle size and morphology remain, necessitating further research to optimise synthesis conditions and expand their potential applications [68].

Unlike algae, actinomycetes are filamentous gram-positive bacteria that produce diverse secondary metabolites, including enzymes and antibiotics. These metabolites play a key role in reducing and stabilising metal ions during nanoparticle synthesis [69]. The process begins with isolating actinomycetes from various natural environments, such as soil or marine habitats, and culturing them under specific conditions [70]. These microorganisms secrete bioactive compounds that act as natural reducing and capping agents, thereby facilitating the formation of nanoparticles. For instance, *Streptomyces tendae* has been used to synthesise silver nanoparticles with strong antimicrobial properties [71], while other actinomycetes have successfully produced magnesium nanoparticles effective against waterborne pathogens [72]. This biosynthetic approach offers multiple advantages, including environmental sustainability by eliminating toxic chemicals, cost-effectiveness through mild operating conditions, and scalability for industrial applications. The resulting nanoparticles have significant applications in medicine, agriculture and environmental remediation.

Regardless of the precursors used for reducing agents, most green synthesis pathways produce nanomaterials with diameters ranging from 1 to 100 nm. Although these sizes are considerably smaller compared with some top-down approaches (e.g. ball mill) and bottom-up approaches (e.g. solvothermal and hydrothermal), the challenges remain. Achieving precise control over the shape and consistency of the nanomaterials is essential [72, 73]. The sizes, shapes and consistency of nanomaterials are key factors that determine their properties and suitability for various applications.

3 Properties of Nanomaterials

Nanoparticles, typically ranging from 1 to 100 nm in diameter [27], possess unique size [74] and surface area characteristics that significantly influence their physical, chemical and biological properties [75]. Their small size results in quantum effects that alter their optical, electrical, chemical and magnetic behaviours. These unique attributes make nanoparticles indispensable in industries ranging from healthcare and cosmetics to energy and environmental technology [76].

Optical Properties

Nanoparticles possess unique optical properties due to quantum effects and surface plasmon resonance (SPR), making them valuable in imaging, biosensing and display technologies [77, 78]. Gold and silver nanoparticles exhibit SPR, leading to strong light absorption and scattering, while quantum dots provide tunable fluorescence based on size, thereby enhancing photodetectors and medical imaging [1]. Similarly, silica-coated gold nanoparticles offer tuneable optical properties for drug delivery and imaging by modifying silica thickness [79].

In addition, CNTs with strong near-infrared absorption are also utilised in photodetectors, solar cells [80] and biomedical therapies. Quantum dots, for instance, exhibit size-dependent fluorescence, emitting different colours based on their size, blue for smaller dots and red for larger ones, making them valuable in imaging, displays and photodetectors [81, 82]. Cadmium selenide (CdSe) quantum dots provide exceptional brightness and stability for high-resolution imaging and display screens [53], while iron oxide nanoparticles featuring both optical and magnetic properties enhance MRI contrast [54]. The field of nanophotonics continues to refine these optical properties, driving advancements in medicine, electronics and energy harvesting. Their ability to interact with light enables innovations in nanophotonics and impacts other characteristics when combined with nanocomposites, such as mechanical strength [17].

Mechanical Properties

In addition to their optical advantages, nanoparticles significantly enhance the mechanical properties of materials. By reinforcing materials, nanoparticles make them stronger, more flexible and more durable. Their high surface area, unique bonding and nanoscale interactions facilitate the development of advanced, high-performance materials. For instance, carbon nanotubes are employed to reinforce composites, resulting in lightweight yet highly durable materials suitable for aerospace and automotive applications [83]. Ceramic nanoparticles like alumina [84] and silicon carbide [85] improve hardness and wear resistance, while silica and nanoclay enhance polymer toughness for packaging and construction [86]. These mechanical improvements are further supported by the chemical stability of nanoparticles, which prevents degradation and extends material lifespan.

Chemical Properties

In addition to mechanical strength, nanoparticles' high surface area and reactivity allow for advanced chemical functionalities [51]. Their ability to act as catalysts [24] and carriers in drug delivery systems demonstrates their impact in industrial [87] and biomedical applications. Platinum nanoparticles enhance fuel cells and catalytic converters [88], while silver nanoparticles provide antimicrobial coatings in medical devices [55]. Iron oxide and cerium oxide nanoparticles contribute to targeted drug delivery and protection against oxidative stress, demonstrating their versatility in both biomedical and environmental applications [89, 90]. Their chemical reactivity also plays a key role in modifying magnetic properties, making them useful in specialised applications.

Magnetic Properties

Magnetic nanoparticles, particularly iron oxide, exhibit superparamagnetism, which is a phenomenon that allows them to be magnetised only in the presence of an external field [91]. These nanoparticles, such as iron oxide, become magnetised when exposed to an external magnetic field, but they do not retain their magnetisation once the field is removed [92]. Magnetic anisotropy enables high-density data storage, as seen in cobalt nanoparticles [93], while giant magnetoresistance (GMR) powers spintronics in hard drives and sensors. Magnetostrictive nanoparticles, which change shape in response to a magnetic field, play a key role in sensors and actuators [94]. Magnetic nanoparticles offer an array of advanced functionalities due to their superparamagnetism, magnetic anisotropy and magnetoresistance [50, 91]. As research progresses, the development of new hybrid magnetic nanomaterials will further expand their applications, paving the way for next-generation medical, electronic and industrial technologies.

Electrical Properties

Nanoparticles revolutionised electrical applications by enhancing conductivity and energy efficiency in electronics. Due to their high surface area and quantum effects, nanoparticles significantly improve electrical properties in various fields, including electronics, optoelectronics and energy storage. CNTs and graphene offer exceptional conductivity and flexibility, making them ideal for nanoelectronics, flexible circuits and high-frequency transistors [33, 80]. Quantum tunnelling in semiconductor nanoparticles, such as quantum dots, enables tunable conductivity, benefiting applications including transistors, displays and memory storage [95]. Silver and copper nanoparticles enable the development of printed electronics and conductive inks, with silver excelling in biosensors and smart textiles [6, 35].

Nanoparticles also improve dielectric properties, with barium titanate enhancing capacitors [96] and indium tin oxide (ITO), enabling transparent conductive films in touchscreens and solar cells [97]. Molybdenum disulfide (MoS_2) nanoparticles, with their tunable bandgap, drive advances in low-power transistors and photodetectors [98]. Nickel and cobalt nanoparticles support magnetoelectronic and spintronics, enhancing data storage [99]. Integrating nanomaterials with electrical properties has revolutionised nanocircuits, wearable electronics and sensors, paving the way for next-generation computing and energy-efficient devices. As research progresses, nanoparticles further transform electronics, energy storage and telecommunications.

4 Conclusions

Nanomaterials have been widely synthesised through various pathways, including top-down methods such as ball milling, lithography, sputtering, laser ablation and arc discharge, as well as bottom-up approaches like CVD, hydrothermal, solvothermal, sol–gel and co-precipitation. Among these, green synthesis has emerged as a highly environmentally friendly and sustainable alternative, producing less hazardous byproducts compared to traditional methods. Nanomaterials have been extensively studied for their unique properties, including high surface area, stability and versatility. These properties make them suitable for diverse applications in fields like aerospace, catalysis, environmental remediation and memory storage. However, achieving precise control over shape and consistency remains a significant challenge. The high processing temperatures and need for sophisticated equipment limit substrate selection and process scalability. Addressing these challenges requires further optimisation and innovation in nanomaterial synthesis techniques. Nanomaterials embedded with metals like silver, gold, copper and cerium have shown great potential in biomedicine, health and environmental remediation. These nanomaterials can lead to advancements in targeted drug delivery, diagnostic imaging and pollution control. The field of nanomaterials continues to make significant contributions to technological innovations and sustainable solutions, driving progress across various sectors.

Acknowledgements The authors thank Universiti Malaysia Sarawak for financial support through the Graduate Research Grant (UNI/F07/GRADUATE/86745/2025).

References

1. P Jiamboonsri S Wanwong 2021 Photoassisted synthesis of silver nanoparticles using riceberry rice extract and their antibacterial application J Nanomater 2021 1 18 https://doi.org/10.1155/2021/5598924
2. Idris DS, Roy A (2023) Synthesis of bimetallic nanoparticles and applications—an updated review. Crystals (Basel) 13:637. https://doi.org/10.3390/cryst13040637
3. I Lacatusu D Istrati N Bordei 2020 Synergism of plant extract and vegetable Oils-based Lipid Nanocarriers: emerging trends in development of advanced cosmetic prototype products Mater Sci Eng, C 108 110412https://doi.org/10.1016/j.msec.2019.110412
4. SF Ahmed M Mofijur N Rafa 2022 Green approaches in synthesising nanomaterials for environmental nanobioremediation: Technological advancements, applications, benefits and challenges Environ Res 204 111967https://doi.org/10.1016/j.envres.2021.111967
5. F Ridwan M Fadel 2021 Preparation of avocado leaf simplicia macro/nano particles by using high energy ball mill IOP Conf Ser Mater Sci Eng 1041 012059https://doi.org/10.1088/1757-899x/1041/1/012059
6. E Sharma R Rathi J Misharwal 2022 Evolution in lithography techniques: microlithography to nanolithography Nanomaterials 12 2754
7. PK Dikshit J Kumar AK Das 2021 Green synthesis of metallic nanoparticles: applications and limitations Catalysts 11 902
8. I Ijaz E Gilani A Nazir A Bukhari 2020 Detail review on chemical, physical and green synthesis, classification, characterizations and applications of nanoparticles Green Chem Lett Rev 13 59 81
9. NS Alharbi NS Alsubhi AI Felimban 2022 Green synthesis of silver nanoparticles using medicinal plants: characterization and application J Radiat Res Appl Sci 15 109 124 https://doi.org/10.1016/j.jrras.2022.06.012
10. J Zhao Y Zhang Y Wang 2018 The application of nanostructured transition metal sulfides as anodes for lithium ion batteries J Energy Chem 27 1536 1554
11. N Salah SS Habib ZH Khan 2011 High-energy ball milling technique for ZnO nanoparticles as antibacterial material Int J Nanomedicine 6 863 869 https://doi.org/10.2147/ijn.s18267
12. Qi Z, Zhao X, Xiuli G, et al (2024) Nano materials preparation mechanism of planetary ball mill. Preprint 1:5001409
13. S Nawaz M Ahmad S Asif 2022 Phyllosilicate derived catalysts for efficient conversion of lignocellulosic derived biomass to biodiesel: a review Bioresour Technol 343 126068
14. ML Menegazzo GG Fonseca 2019 Biomass recovery and lipid extraction processes for microalgae biofuels production: a review Renew Sustain Energy Rev 107 87 107
15. Mohamed EH, Ebeid HM, Ashoush IS et al (2023) Nano-capsulation of ginger, red cabbage and broccoli ball mill extracts as sources of anti-oxidant and anti-cancer and application in lentil soup powder. Egypt J Chem 66:361–379. https://doi.org/10.21608/EJCHEM.2022.154781.6692

16. N Al-Harbi NK Abd-Elrahman 2024 Physical methods for preparation of nanomaterials, their characterization and applications: a review J Umm Al-Qura Univ Appl Sci 2024 43994
17. W Liu J Wang X Xu 2021 Single-step dual-layer photolithography for tunable and scalable nanopatterning ACS Nano 15 12180 12188 https://doi.org/10.1021/acsnano.1c03703
18. Gupta A (2023) Review of: Amplification of nano wires nano wire by electron nano lithography. Qeios 4:1. https://doi.org/10.32388/l3md1n
19. M Li Y Chen W Luo X Cheng 2021 Interfacial interactions during demolding in nanoimprint lithography Micromachines (Basel) 12 349 https://doi.org/10.3390/mi12040349
20. R Maheswaran BP Shanmugavel 2022 A critical review of the role of carbon nanotubes in the progress of next-generation electronic applications J Electron Mater 51 2786 2800
21. Orozco-Montes V, Caillard A, Brault P et al (2021) Synthesis of platinum nanoparticles by plasma sputtering onto glycerol: effect of argon pressure on their physicochemical properties. J Phys Chem C 2021:3169–3179. https://doi.org/10.1021/acs.jpcc.0c09746ï
22. Abaullah M (2023) Preparation of silver nano-particles in benzene solvent by laser ablation method. J Duhok Univ 26:139–149. https://doi.org/10.26682/sjuod.2023.26.1.14
23. A Sergievskaya A Chauvin S Konstantinidis 2022 Sputtering onto liquids: a critical review Beilstein J Nanotechnol 13 10 53 https://doi.org/10.3762/BJNANO.13.2
24. J Yu X Sun X Tong 2021 Ultra-high thermal stability of sputtering reconstructed Cu-based catalysts Nat Commun 12 7209 https://doi.org/10.1038/s41467-021-27557-1
25. D Valerini L Tammaro R Vitali 2021 Sputter-deposited ag nanoparticles on electrospun pcl scaffolds: morphology, wettability and antibacterial activity Coatings 11 345 https://doi.org/10.3390/coatings11030345
26. Bokov D, Turki Jalil A, Chupradit S et al (2021) Nanomaterial by sol-gel method: synthesis and application. Adv Mater Sci Eng 2021:1–21
27. N Baig I Kammakakam W Falath I Kammakakam 2021 Nanomaterials: a review of synthesis methods, properties, recent progress, and challenges Mater Adv 2 1821 1871
28. Alwazny MS, Ismail RA, Salim ET (2022) Aggregation threshold for novel Au-$LiNbO_3$ core/shell nano composite: effect of laser ablation energy fluence. 15:2022
29. Y Khan H Sadia SZ Ali Shah 2022 Classification, synthetic, and characterization approaches to nanoparticles, and their applications in various fields of nanotechnology: a review Catalysts 12 1386
30. Kalai Priya A, Yogesh GK, Subha K et al (2021) Synthesis of silver nano-butterfly park by using laser ablation of aqueous salt for gas sensing application. Appl Phys A Mater Sci Process 127:292. https://doi.org/10.1007/s00339-021-04370-7
31. Z Jiang L Li H Huang 2022 Progress in laser ablation and biological synthesis processes: "top-down" and "bottom-up" approaches for the green synthesis of Au/Ag nanoparticles Int J Mol Sci 23 14658
32. V Harish MM Ansari D Tewari 2022 Nanoparticle and nanostructure synthesis and controlled growth methods Nanomaterials 12 3226
33. Tepale-Cortés A, Moreno-Saavedra H, Hernandez-Tenorio C et al (2021) Multi-walled carbon nanotubes synthesis by arc discharge method in a glass chamber. J Mex Chem Soc 65:480–490. https://doi.org/10.29356/JMCS.V65I4.1486
34. Stetsenko M, Abaszade R (2023) X-ray phase analysis of carbon nanotubes obtained by the arc discharge method. UNEC J Eng Appl Sci 3:15–20. https://doi.org/10.61640/ujeas.2023.0503
35. HM Dewey A Lamb J Budhathoki-Uprety 2024 Recent advances on applications of single-walled carbon nanotubes as cutting-edge optical nanosensors for biosensing technologies Nanoscale 16 16344 16375
36. K Ramachandran V Boopalan JC Bear R Subramani 2022 Multi-walled carbon nanotubes (MWCNTs)-reinforced ceramic nanocomposites for aerospace applications: a review J Mater Sci 57 3923 3953
37. AM El-Khatib MS Badawi ZF Ghatass 2018 Synthesize of silver nanoparticles by arc discharge method using two different rotational electrode shapes J Clust Sci 29 1169 1175 https://doi.org/10.1007/s10876-018-1430-2

38. Ashkarran AA, Iraji zad A, Mahdavi SM et al (2009) Rapid and efficient synthesis of colloidal gold nanoparticles by arc discharge method. Appl Phys A Mater Sci Process 96:423–428. https://doi.org/10.1007/s00339-009-5288-x
39. L Sun G Yuan L Gao 2021 Chemical vapour deposition Nat Rev Methods Prim 1 5
40. L Seravalli M Bosi 2021 A review on chemical vapour deposition of two-dimensional mos2 flakes Materials 14 7590
41. N Nandihalli DH Gregory T Mori 2022 Energy-saving pathways for thermoelectric nanomaterial synthesis: hydrothermal/solvothermal, microwave-assisted, solution-based, and powder processing Adv Sci 9 2106052
42. Singh P, Mandal D, Chandra A, Singh T (2024) Zinc stannate oxide perovskite nanomaterial based electrochemical detection of ammonia. 366:114955
43. F Riporto A Dhouib A Gheata 2024 Nonclassical nucleation and crystallization of $LiNbO_3$ nanoparticles from the aqueous solvothermal alkoxide route Small 20 2306417 https://doi.org/10.1002/smll.202306417
44. ND Jaji MBH Othman HL Lee 2021 One-pot solvothermal synthesis and characterization of highly stable nickel nanoparticles Nanotechnol Rev 10 318 329 https://doi.org/10.1515/ntrev-2021-0019
45. Saremi A, Mirkazemi SM, Sazvar A, Rezaie H (2024) Controlling magnetic and surface properties of cobalt ferrite nanoparticles: a comparison of co-precipitation and solvothermal synthesis methods. 148:107432
46. L Marinescu D Ficai A Ficai 2022 Comparative antimicrobial activity of silver nanoparticles obtained by wet chemical reduction and solvothermal methods Int J Mol Sci 23 5982 https://doi.org/10.3390/ijms23115982
47. MDI Bhuyan S Das MA Basith 2021 Sol-gel synthesized double perovskite Gd_2FeCrO_6 nanoparticles: structural, magnetic and optical properties J Alloys Compd 878 160389https://doi.org/10.1016/j.jallcom.2021.160389
48. GH Patel SH Chaki RM Kannaujiya 2021 Sol-gel synthesis and thermal characterization of SnO_2 nanoparticles Physica B Condens Matter 613 412987https://doi.org/10.1016/j.physb.2021.412987
49. DA Patiño-Ruiz SI Meramo-Hurtado ÁD González-Delgado A Herrera 2021 Environmental Sustainability evaluation of iron oxide nanoparticles synthesized via green synthesis and the coprecipitation method: a comparative life cycle assessment study ACS Omega 6 12410 12423 https://doi.org/10.1021/acsomega.0c05246
50. Z Li Y Sun S Ge 2023 An overview of synthesis and structural regulation of magnetic nanomaterials prepared by chemical coprecipitation Metals (Basel) 13 152 https://doi.org/10.3390/met13010152
51. Deng N, Li Z, Zuo X et al (2021) Coprecipitation of Fe/Cr hydroxides with organics: roles of organic properties in composition and stability of the coprecipitates. 55:4638–4647
52. Khan MS, Ranjani S, Hemalatha S (2022) Synthesis and characterization of Kappaphycus alvarezii derived silver nanoparticles and determination of antibacterial activity. Mater Chem Phys 282:125985. https://doi.org/10.1016/j.matchemphys.2022.125985
53. K Hachem MJ Ansari RO Saleh 2022 Methods of chemical synthesis in the synthesis of nanomaterial and nanoparticles by the chemical deposition method: a review Bionanoscience 12 1032 1057
54. N Priya K Kaur AK Sidhu 2021 Green synthesis: an eco-friendly route for the synthesis of iron oxide nanoparticles Front Nanotechnol 3 655062
55. Murshedi LDN, Solyanikova IP (2024) The green synthesis and phytochemical properties of silver nanoparticles obtained from eggplant. In: The 3rd international electronic conference on processes. MDPI, Basel Switzerland, p 72
56. M Huston M Debella M Dibella A Gupta 2021 Green synthesis of nanomaterials Nanomaterials 11 2130
57. IK Siakavella F Lamari D Papoulis 2020 Effect of plant extracts on the characteristics of silver nanoparticles for topical application Pharmaceutics 12 1 17 https://doi.org/10.3390/pharmaceutics12121244

58. VP Aswathi S Meera CGA Maria M Nidhin 2023 Green synthesis of nanoparticles from biodegradable waste extracts and their applications: a critical review Nanotechnol Environ Eng 8 377 397
59. T Gur I Meydan H Seckin 2022 Green synthesis, characterization and bioactivity of biogenic zinc oxide nanoparticles Environ Res 204 111897https://doi.org/10.1016/j.envres.2021.111897
60. Ahmad A, Javed MS, Khan S et al (2023) Green synthesized Ag decorated CeO_2 nanoparticles: Efficient photocatalysts and potential antibacterial agents. Chemosphere 310. https://doi.org/10.1016/j.chemosphere.2022.136841
61. Abeer Mohammed AB, Abd Elhamid MM, Khalil MKM et al (2022) The potential activity of biosynthesized silver nanoparticles of Pseudomonas aeruginosa as an antibacterial agent against multidrug-resistant isolates from intensive care unit and anticancer agent. Environ Sci Eur 34:109. https://doi.org/10.1186/s12302-022-00684-2
62. C Campo-Beleño RA Villamizar-Gallardo LE López-Jácome 2022 Biologically synthesized silver nanoparticles as potent antibacterial effective against multidrug-resistant Pseudomonas aeruginosa Lett Appl Microbiol 75 680 688 https://doi.org/10.1111/lam.13759
63. Y Rilda DN Syafitri U Septiani 2024 A comparative analysis of capping and reducing agents of microbial cell Aspergillus niger and Bacillus subtilis for biosynthesis of Ag doped ZnO nanoparticles Ceram Int 50 28150 28158 https://doi.org/10.1016/j.ceramint.2024.05.114
64. P Nag K Sadani L Pisharody 2024 Essential oil mediated synthesis and application of highly stable copper nanoparticles as coatings on textiles and surfaces for rapid and sustained disinfection of microorganisms Nanotechnology 35 345602https://doi.org/10.1088/1361-6528/ad501b
65. A Mukherjee D Sarkar S Sasmal 2021 A review of green synthesis of metal nanoparticles using algae Front Microbiol 12 693899
66. R He D Zhou L Xiao Y Li 2023 Chlorella vulgaris extract-decorated gold nanoparticle hybridized antimicrobial hydrogel as a potential dressing Gels 9 11 https://doi.org/10.3390/gels9010011
67. R Thiurunavukkarau S Shanmugam K Subramanian 2022 Silver nanoparticles synthesized from the seaweed Sargassum polycystum and screening for their biological potential Sci Rep 12 14757 https://doi.org/10.1038/s41598-022-18379-2
68. R Chaudhary K Nawaz AK Khan 2020 An overview of the algae-mediated biosynthesis of nanoparticles and their biomedical applications Biomolecules 10 1 35
69. MS Anjum S Khaliq N Ashraf 2024 Bioactive streptomycetes: a powerful tool to synthesize diverse nanoparticles with multifarious properties J Basic Microbiol 64 1 15
70. S Aati HY Aati AA Hamed 2025 Gold nanoparticles synthesized from soil-derived *Streptomyces* sp. ASM19: characterization, antimicrobial, anticancer potency targeted G2/M phase cell-cycle arrest, and *in silico* studies RSC Adv 15 3954 3968 https://doi.org/10.1039/D4RA07608G
71. A Bizuye L Gedamu C Bii 2020 Molecular-based identification of actinomycetes species that synthesize antibacterial silver nanoparticles Int J Microbiol 2020 1 17 https://doi.org/10.1155/2020/8816111
72. VV Chougule S Bangale 2021 Green synthesis of magnesium nanoparticles from actinomycetes and their therapeutic applications J Adv Sci Res 12 276 284
73. C Hano BH Abbasi 2022 Plant-based green synthesis of nanoparticles: production, characterization and applications Biomolecules 12 31 https://doi.org/10.3390/biom12010031
74. Rana A, Yadav K, Jagadevan S (2020) A comprehensive review on green synthesis of nature-inspired metal nanoparticles: mechanism, application and toxicity. J Clean Prod 272
75. K Niska E Zielinska MW Radomski I Inkielewicz-Stepniak 2018 Metal nanoparticles in dermatology and cosmetology: interactions with human skin cells Chem Biol Interact 295 38 51 https://doi.org/10.1016/j.cbi.2017.06.018
76. Souto EB, Fernandes AR, Martins-Gomes C et al (2020) Nanomaterials for skin delivery of cosmeceuticals and pharmaceuticals. Appl Sci (Switzerland) 10
77. E Koushki 2021 Effect of conjugation with organic molecules on the surface plasmon resonance of gold nanoparticles and application in optical biosensing RSC Adv 11 23390 23399 https://doi.org/10.1039/d1ra01842f

78. MS Rodrigues J Borges C Lopes 2021 Gas sensors based on localized surface plasmon resonances: synthesis of oxide films with embedded metal nanoparticles, theory and simulation, and sensitivity enhancement strategies Appl Sci (Switzerland) 11 5388 https://doi.org/10.3390/app11125388
79. MI Anik N Mahmud A Al Masud M Hasan 2022 Gold nanoparticles (GNPs) in biomedical and clinical applications: a review Nano Select 3 792 828 https://doi.org/10.1002/nano.202100255
80. E Muchuweni ET Mombeshora BS Martincigh VO Nyamori 2022 Recent applications of carbon nanotubes in organic solar cells Front Chem 9 733552
81. J Fan L Qian 2022 Quantum dot patterning by direct photolithography Nat Nanotechnol 17 906 907
82. Kasinathan K, Samayanan S, Marimuthu K, Yim JH (2022) Green synthesis of multicolour fluorescence carbon quantum dots from sugarcane waste: investigation of mercury (II) ion sensing, and bio-imaging applications. Appl Surf Sci 601. https://doi.org/10.1016/j.apsusc.2022.154266
83. A Nag MEE Alahi SC Mukhopadhyay Z Liu 2021 Multi-walled carbon nanotubes-based sensors for strain sensing applications Sensors (Switzerland) 21 1 22
84. BV Murthy V Auradi M Nagaral 2023 Al2014-Alumina aerospace composites: particle size impacts on microstructure, mechanical, fractography, and wear characteristics ACS Omega 8 13444 13455 https://doi.org/10.1021/acsomega.3c01163
85. SM Shahabaz PK Shetty N Shetty 2022 Effect of alumina and silicon carbide nanoparticle-infused polymer matrix on mechanical properties of unidirectional carbon fiber-reinforced polymer J Compos Sci 6 381 https://doi.org/10.3390/jcs6120381
86. C Barros S Miranda O Castro 2023 LDPE-Nanoclay films for food packaging with improved barrier properties J Plast Film Sheeting 39 304 320
87. Kheiriabad S, Jafari A, Namvar Aghdash S, et al (2024) Applications of advanced nanomaterials in biomedicine, pharmaceuticals, agriculture, and food industry. Bionanoscience 4298–4321. https://doi.org/10.1007/s12668-024-01506-w
88. S Navarro-Espinoza D Meza-Figueroa R Guzmán 2021 Release of nanoparticles in the environment and catalytic converters ageing Nanomaterials 11 3406 https://doi.org/10.3390/nano11123406
89. A Saha K Giri SP Agarwal 2017 Silver nanoparticles based hydrogels of tulsi extracts for topical drug delivery Int J Ayurveda Pharma Res 5 17 23
90. K Hkiri H Elsayed Ahmed Mohamed S Ghotekar M Maaza 2024 Green synthesis of cerium oxide nanoparticles using Portulaca oleracea Extract: photocatalytic activities Inorg Chem Commun 162 112243https://doi.org/10.1016/j.inoche.2024.112243
91. A Surpi T Shelyakova M Murgia 2023 Versatile magnetic configuration for the control and manipulation of superparamagnetic nanoparticles Sci Rep 13 5301 https://doi.org/10.1038/s41598-023-32299-9
92. Pucci C, Degl'Innocenti A, Belenli Gümüş M, Ciofani G (2022) Superparamagnetic iron oxide nanoparticles for magnetic hyperthermia: recent advancements, molecular effects, and future directions in the omics era. Biomater Sci 10:2103–2121
93. A Zorai A Souici D Dragoe 2022 Superparamagnetic cobalt ferrite nanoparticles synthesized by gamma irradiation New J Chem 47 2626 2634 https://doi.org/10.1039/d2nj05433g
94. S Yang J Zhang 2021 Current progress of magnetoresistance sensors Chemosensors 9 211
95. SY Li L He 2022 Recent progresses of quantum confinement in graphene quantum dots Front Phys (Beijing) 17 33201 https://doi.org/10.1007/s11467-021-1125-2
96. Tewatia K, Sharma A, Sharma M, Kumar A (2020) Factors affecting morphological and electrical properties of Barium Titanate: a brief review. In: Materials today: proceedings. Elsevier Ltd, pp 4548–4556
97. QM Al-Bataineh AB Migdadi A Telfah 2022 Physical and chemical characterization of polyaniline (PANI)/indium tin oxide nanoparticles (ITONPs) nanocomposite films Mater Chem Phys 290 126387https://doi.org/10.1016/j.matchemphys.2022.126387
98. P Luo C Liu J Lin 2022 Molybdenum disulfide transistors with enlarged van der Waals gaps at their dielectric interface via oxygen accumulation Nat Electron 5 849 858 https://doi.org/10.1038/s41928-022-00877-w

99. MH Khatun R Amin MSI Sarker 2024 Structural and magnetic properties of Fe and Ni co-doped SnO_2 nanoparticles prepared by co-precipitation method Mater Res Express 11 16102 https://doi.org/10.1088/2053-1591/ad1d87

Graphene Nanomaterials for Separation of Hazardous Materials from Wastewater

Ravi Kumar

Abstract Water pollution has raised the alarms toward the maintenance of life on earth. Industrial revolution has given a big blow to the aquatic health of the planet. Treatment of wastewater is one of the foremost issues being considered by the scientific community. Carbon-based nanomaterials have grabbed the interest of the research community owing to their noteworthy characteristics. These nanomaterials include carbon nanotubes, graphene, graphene oxide, reduced graphene oxide. Graphene-based nanomaterials are found to have different roles in wastewater treatment via processes like adsorption, catalysis, and membrane filtration owing to their remarkable properties and the different mechanisms associated with their activity. This chapter provides a useful insight into the various roles and the factors associated with the appliance of graphene-based nanomaterials for wastewater treatment along with their pros and cons—a future-looking perspective.

Keywords Graphene · Graphene oxide · Adsorption · Photocatalysis · Membrane filtration technology

1 Introduction

Water is among the most fundamental and indispensable resources for the sustenance of life on Earth [1]. Water covers nearly 70% of earth's surface and it is the only planet that is livable [2]. Despite the vast amount of water on Earth, only 2.5% is freshwater suitable for consumption and essential ecological processes, while the remaining 97.5% consists of saline water found primarily in seas and oceans [3, 4]. Major problems with water management include the rise in water consumption and the contamination of the water supply [5]. Freshwater is contaminated by various human activities, rendering it unsuitable for domestic and agricultural use. In reference to the risk of a global water crisis, the journal *Nature* highlights that "conditions

R. Kumar (✉)
Department of Chemistry and Center for Advanced Studies, Panjab University, Chandigarh, India
e-mail: ravikumar2167@gmail.com

S. A. Bhawani and M. Jawaid (eds.), *Nanomaterials for Separation of Hazardous Contaminants from Wastewater*, Smart Nanomaterials Technology,
https://doi.org/10.1007/978-981-95-6079-0_2

are worsening, and over one trillion people lack access to clean water" [6]. The issue is worsened by the disposal of electronic waste from industries and large-scale industrialization. Hazardous metals like Lead (Pb), Mercury (Hg), Arsenic (As), Cobalt (Co), Nickel (Ni), and other environmentally harmful substances are being introduced into water sources [7]. Due to severe contamination of water, it has become a global concern to assure the availability of water that is safe for drinking purposes [8]. Emergent contaminants can also be hazardous to environments and human health at low concentrations [9].

In addition to more sophisticated techniques (ultra-filtration, chemical precipitation, adsorption, biological oxidation, coagulation, biosensor, and micro screening), there are a number of traditional approaches (coagulation, electrochemical treatment, ion-exchange, adsorption, membrane filtration, and chemical precipitation) for treating contaminated water [10]. In addition to being expensive, typical water treatment methods are ineffective. Toxic heavy metals are typically found in industrial wastewater [11]. As a result, the wastewater treatment necessitates the implementation of an energy- and cost-efficient technology.

Carbon-based substances, primarily Activated Carbon (AC), have extensively been studied as excellent adsorbents for eliminating different organic and inorganic pollutants [12, 13]. These nanomaterials (NMs) are realized in the form of AC, graphene (GR), fullerenes, and carbon nanotubes (CNTs), etc. These features include high porosity, adjustable pore size, and high surface area combined with remarkable electronic conductivities as well as chemical stability in both acid/alkaline electrolytic environments [14].

GR, a single layer of graphite, commonly utilized in both fundamental and applied scientific research [15, 16]. Recently, GR and graphene oxide (GO) NMs have found successful applications across various industries, including sensing, biology, medicine, and catalysis. These NMs, in particular, have acquired consideration as a method for cleaning wastewater. It is because of the high specific surface area that ensures high adsorption capacity of these materials. Additionally, these nanoparticles (NPs) demonstrate a high degree of mobility in solutions. Numerous studies have shown that GR and GO NPs are effective at removal of microorganisms, heavy metal pollutants, and both organic and inorganic anions. Research indicates that GR and GO NPs hold significant potential for treating both wastewater and drinking water.

Among these, GR NPs and nanocomposites are currently the focus of extensive studies toward wastewater treatment. GR is the best adsorbent for the elimination of different antibiotics through π–π interactions. Though, aggregation in aqueous solution poses the biggest problem for it. This constraint limits both the material's ability to adsorb and be recycled. Functionalization is one method for overcoming this restriction [17]. According to Puri et al. [18], GO, a chemically functionalized derivative of GR, demonstrates exceptional hydrophilicity, making it highly suitable for adsorption applications. The presence of hydrophilic carboxyl and hydroxyl (OH) functional groups on the surface of GO nanosheets significantly enhances their capacity to adsorb Heavy Metal Ions (HMIs) from wastewater [19]. Organic water contaminants are adsorbed on GO through conventional forces like coulombic forces, H-bonding, or π–π interactions. In contrast, inorganic metal ions are adsorbed onto

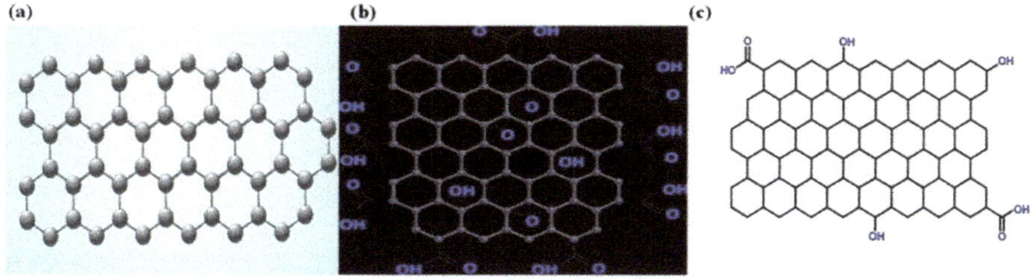

Fig. 1 Structures of various GR analogs: **a** Gr, **b** GO, and **c** rGO

GO nanosheets via Lewis acid–base interactions and chemisorption. This dual functionality enables these carbonaceous nanosheets to effectively remove both metal and organic pollutants from water. Furthermore, surface engineering can chemically modify these nanosheets by incorporating NPs or specific hydrophilic functional groups, which enhances their catalytic adsorption or degradation capabilities for water contaminants [20]. Figure 1 shows the various forms of GR-based NMs applied for the wastewater treatment. The critical physical properties of materials derived from GR include high surface area, tunable pore size, and porosity. In contrast, key chemical properties encompass purity, surface chemistry, surface charge, and polarity. A larger surface area enhances the performance of adsorption systems. However, the pore size and volume of the GR must correspond to the size of the contaminants; for example, a small pore size will hinder the adsorption of the hazardous. Research indicates that GR contains meso- and macropores, while GO features meso- and micropores. Therefore, it is essential to explore these differences to design efficient wastewater treatment methods utilizing GR-based NMs [21]. The above-discussed fascinating properties of the GR-based NMs make them perfect candidates for the wastewater treatment with different roles in the process, which are discussed in the upcoming sections of the chapter.

1.1 Different Roles of Graphene-Based Nanomaterials for Wastewater Treatment

GR, GO, and rGO have grabbed the attention as a solution for the rising concern of the wastewater. Owing to their different properties, different functional groups (mainly Oxygen-rich Functional Groups (OFGs)), tunable pore size and large surface area, they involve different kind of mechanisms for the wastewater treatment, depending upon the nature of the pollutant, i.e., whether it is an organic pollutant or inorganic pollutant. These materials can play various roles in the process like adsorbent, catalyst, or the filtration membrane (see Fig. 2) which are discussed in detail in the following parts of the chapter.

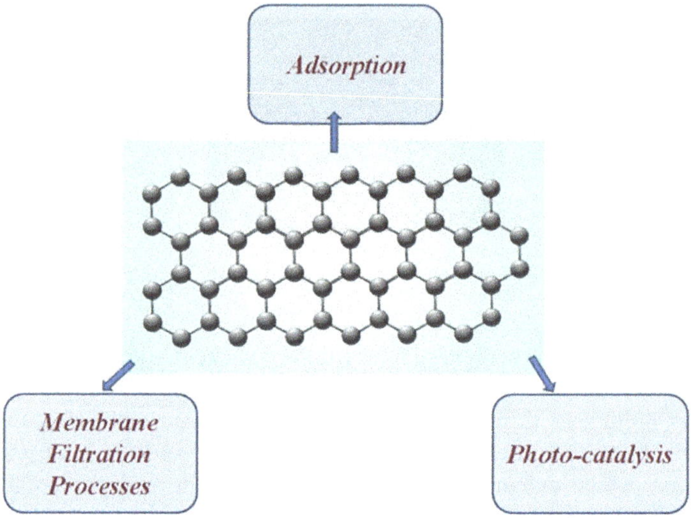

Fig. 2 Different roles of GR in wastewater treatment process

2 Graphene-Based Nanomaterials as Adsorbent

Adsorption is regarded as an effective and straightforward wastewater treatment method for elimination of different contaminants from water sources, including organic dyes, HMIs, phenols, and others [22]. The effectiveness of an adsorption process in separating materials is primarily determined by the adsorbent's capacity to adsorb. Specifically, the amount of active adsorption sites and the exposed surface area of adsorbent are the critical factors influencing contaminant adsorption. Due to high specific area and effective adsorption sites, NPs emerge as exceptional adsorbents. Among these, carbon-based NMs are the most commonly utilized adsorbents, as they are widely available, exhibit strong adsorption capabilities, possess high chemical and thermal stability, and can be easily sourced.

The adsorption characteristics of GR-based materials are significantly affected by their surface chemistry. Functional groups containing oxygen exert two competing influences on the adsorption capacity of GO surfaces. While these groups can enhance water solubility on the surface, thereby improving adsorption, they may also result in reduction of available adsorption sites because of the formation of water clusters on the surface. Conversely, availability of OFGs on GO can facilitate the adsorption of contaminants, like hydroxyl compounds or amino acids via strong H-bonding or Lewis interactions. To ensure accurate predictions of treatment performance, it is essential that GR-based materials maintain a high level of purity [22].

2.1 Adsorption of Organic Contaminants

Bacteria and other organic pollutants found in water can originate from various sources, including sewage, fertilizers, agricultural runoff, forestry operations, food processing, debris from trees and brush, industrial waste, and other contributors. Pure sp^2 hybridized carbon is hydrophobic and promotes efficient adsorption of hydrocarbons due to its surface characteristics, which are characterized by negative interactions. However, it remains partially reactive because of the presence of dangling bonds or the diverse modifications of GR. This material is effective at absorbing organic compounds with both hydrophilic and hydrophobic characteristics, indicating that π–π interactions remain strong even in the presence of functional groups. Among organic contaminants, key attributes include geometry, size, functional groups, hydrophobicity, and aromaticity. The hydrophobicity of organic contaminants is a crucial factor in adsorption systems [23]. GR exhibits compatibility with hydrophobic or polar organic pollutants, whereas it is incompatible with ionic or acidic/basic contaminants. The adsorption capability of organic pollutants is greatly influenced by the quantity and spatial arrangement of aromatic rings. Studies show that a molecule's propensity for adsorption is increased when its structure has more aromatic rings [24]. The application of GR-based adsorbents has become gradually more critical because of accumulation of organic pollutants in the human body, which poses toxic risks and threatens both the environment and human health. Two main aspects affect how these contaminants are treated: the system's adsorption capabilities and the qualities of the materials utilized [25]. In order to get a deeper perceptive of the adsorption properties of GR, it is essential to identify the affinities and interactions between GR and organic contaminants. The treatment of GR wastewater involves a number of interactions, including Lewis' acid–base, electrostatic, van der Waals, hydrophobic, electron donor–acceptor, and H-bonding interactions. The performance of these systems is also significantly influenced by GR materials, organic contaminants, and background solutions [26]. Jiang et al. [27] investigated the 17b-Estradiol (E2) (estrogen steroid hormone) adsorption process on a few layered GO nanosheets for wastewater treatment. The research found that the adsorption of E2 on GO nanosheets is primarily attributed to strong interactions, specifically π–π interactions and H-bonding. This is because of the presence of hydrophilic OH and carboxyl group on the surfaces of both GO and E2. Another paper used a single-step pyrolysis to create GR sand hybrid utilizing date syrup (D-GSH) for wastewater purification [28]. It is found that the hydrophilic surface of the desert sand changes from hydrophilic to hydrophobic after hydrophobic GR is immobilized. This assembly of hybrid D-GSH, that has negatively charged groups and a conjugated structure of π–electrons, makes it easier to remove metal ions (MIs), anionic and cationic dyes, and both via potential adsorption mechanisms. This cost-effective and environmentally friendly technology acted as an efficient metal scavenger for wastewater treatment. D-GSH demonstrated impressive adsorption capacities of 2564 mg/g for Methyl Violet (MV) dye, 333 mg/g for Congo

Red (CR) dye, 781 mg/g for Pb(II), and 793 mg/g for Cd(II) contaminants from wastewater.

Organic contaminants, Polycyclic Aromatic Hydrocarbons (PAHs) are mostly created when coal, oil, and gas are not entirely burnt. Because of the presence of aromatic pollutants like PAHs, industrial wastewater, especially that from the oil industry, has a large potential for pollution. This has led to the application of several technologies for treating industrial effluents in an effort to address socio-environmental and economic needs by removing such toxins from ecosystems. The presence of PAHs in wastewater is a big reason to worry for the industries because of the possible harm that these chemicals can do to persons and the nature. Scientists have become interested in some GR-based compounds because they can be used to remove PAHs from wastewater. The utilization of GR-based materials for treatment of effluents through adsorption processes is attributed to unique properties and low production cost of these materials [29]. A GR-grafted material with 9-octadecenoic acid was created by Alghunaimi et al. and used to remove organic pollutants from oily water. By chemically treating graphite, the acid was attached to GR to create this 9-octadecenoic acid grafted GR (OG). POG, a hydrophobic substance, was then produced via emulsion polymerization of styrene. The study assessed how well POG composites separated organic pollutants such as alkanes having (6, 7, 9, 10, and 11) carbon atoms from water. The findings showed that as the amount of GR grafted with 9-octadecenoic acid increased, so did the adsorption efficiency of the composite and the rate of organic component adsorption. Alvarez et al. [30] evaluated the effectiveness of using GO, synthesized using the traditional Hummer's method, for the adsorption of petrogenic PAHs from water resources. TEM and AFM techniques were used to characterize the materials that were obtained. The maximal platelet length was determined by TEM examination to be around 13 μm, while the GO platelets' thickness was determined by AFM investigation to be 0.612 0.176 nm. According to the adsorption data, BaP has a high GO adsorption capacity because a 98% reduction was attained after 40 h for an initial concentration of 100 g/L. After 40 h, there was a 95.7% reduction for PHEN, an 84.4% reduction for FLU, and a 51.5% reduction for ACE for the other PAH components. Song et al. employed a hydrothermal method for synthesizing N-doped rGO (NRGO). The resulting material was used as an adsorbent to remove ANT and 2-MAQ from synthetic wastewater. The presence of the functional groups in the GO, rGO, and NRGO is confirmed by the FTIR analysis. The addition of nitrogen to rGO's structure resulted in reduction of the folding of crushed leaves during the thermal reduction process, while the overall GR structure remained intact. SEM analysis showed that NRGO exhibited a lower degree of layer aggregation compared to rGO. The adsorption capacities for ANT and 2-MAQ were found to be 5.77 mg/g and 9.29 mg/g, respectively. This study demonstrated that reduction of GO and subsequently incorporating nitrogen can enhance its properties and improve the binding capacity for hydrophobic compounds like PAHs [31]. Dyes are another class of organic compounds which cause serious threat toward the health of aquatic environment, needs to be removed from the aquatic bodies. In order to adsorb the dye Malachite Green (MG) from wastewater, Chen et al. developed a GO/aminated lignin aerogel (GO/AL-AG) and investigated its adsorption capacity across various dosages, pH levels, contact times, and reaction temperatures. The experiments revealed that under optimal conditions, the prepared GO/AL-AG exhibited the highest adsorption capacity of 113.5 mg/g, and an efficiency of 91.72%. The adsorption efficiency of GO/AL-AG considerably outperformed other aerogels, and was attributed to the interaction between the carboxyl groups present on the surface of the GO, and the amino group of the aminated lignin. After five cycles of adsorption–desorption, GO/AL-AG retained nearly 90% of its adsorption effectiveness [32]. To remove Methylene Blue (MB) from wastewater efficiently, Bu et al. synthesized GO functionalized with thiosemicarbazide (TSC) (GO-TSC). The GO and GO-TSC exhibited the maximum adsorption capacities of 196.8 mg/g and an impressive 596.642 mg/g, respectively, demonstrating remarkable stability [33]. Zheng and colleagues developed a hydrogel composed of GO coated with silver NPs (AgNPs) and subsequently integrated it with a porphyrin complex. Their findings revealed that the tetraphenylporphyrin-modified hydrogel showed the highest adsorption capacity for MB, reaching an impressive 130.37 mg/g after treatment with various porphyrin complexes [34]. Sirajudheen and his colleagues recently developed a new chitosan-supported GO hydro-composite (GO/CS-HCP), and it was found to be very efficient toward removal of organic contaminants from polluted water. In particular, it showed excellent adsorption capacities for CR, Acid Red 1 (AR1), and Reactive Red 2 (RR2) dyes. The adsorption was highest at pH 2 and decreased as the pH increased. The GO/CS-HCP also showed good regeneration ability, with a desorption efficiency of over 65% in 0.1 M NaOH solution. All things considered, the GO/CS-HCP is a promising novel material for wastewater organic pollution treatment [35]. Similar results were found for the cationic dye MG, where rGO had a reasonably high adsorption capacity (472.6 mg/g) [36, 37]. The high surface area of rGO played a crucial role in efficiently removing malachite green dye from wastewater. The aromatic rings of dye interacted with rGO's hexagonal carbon lattice, while residual oxygen and electrostatic attraction between the dye's cationic center and rGO's electron clouds further enhanced removal. Despite its effectiveness, further research is necessary to optimize rGO's performance for removing cationic dyes from wastewater [36].

2.2 Adsorption of Inorganic Pollutants from Wastewater

The term "inorganic water pollutants" refers to a broad category of materials, such as metals, acids, salts, and their compounds. Numerous things, including mining operations, volcanic eruptions, and industrial processes, can produce these pollutants. One major environmental worry is acid rain, which is caused by emissions of oxides of sulfur and nitrogen. Because of its wastewater discharges, the chemical sector is a significant source of inorganic water pollution.

In aqueous solutions, heavy metals become enriched and become very hard to decompose. MIs must first be detected in order to be removed, but there aren't any practical applications for sensitive and reliable ways to do this [38]. MIs may now be more accurately detected using sensors based on rGO. The designs employ more OFGs on the surface of rGO that is directly related to their capacity for MI detection. rGO-based nanocomposites were initially designed for the recognition of a single MI, but more recent research has shifted its focus to the creation of sensors that could concurrently sense numerous MIs. To increase sensitivity for multi-ion detection, several of these methods significantly modified rGO [39]. Furfural/rGO composites were used to create a sensor that can detect different HMIs. Because of its naturally high specific surface area, the sensor integrated a significant number of OFGs as heavy metal binding sites. In real-world samples, the furfural/rGO composite sensor achieved a respectable recovery rate of 93.3–105.8% by successfully detecting Cd^{2+}, Pb^{2+}, Cu^{2+}, and Hg^{2+} ions both alone and in combination. The interaction between the metal cations and the hydrophilic groups on the sensor allows for the efficient detection of HMIs [40]. GR has been demonstrated to be an efficient adsorbent for the elimination of Fe(II), fluoride (F^-), and Co(II). Furthermore, GO is a great adsorbent for a variety of cationic contaminants, such as metals like Pd, Cu, Ni, and Zn. OFGs on GO sheets serve as anchors during the adsorption process, improving their capacity to efficiently absorb these metals [29]. An effective method for separating inorganic contaminants is said to be covering GR and GO with magnetic NPs like magnetite [41, 42]. Highly toxic rare earth MIs are linked to severe water contamination. GO has proven effective in removing La, Gd, Y, and Nd from wastewater. Additionally, GO nanocomposites functionalized with Polyaniline (PANI), magnetite, and titanium phosphate have been proposed as promising alternatives for this application [43]. RGO/$Mg(OH)_2$–MgO nanocomposites were used by Purwajanti et al. to show how to adsorb As(III) from domestic water. The in-situ production of $Mg(OH)_2$ nanosheets and improving the accessibility of sorption sites for As(III) depend on the optimized composition of these nanocomposites (MgO = 0.5 and reduced crystallinity). The analyzed nanocomposites had the largest sorption capacity, which is consistent with a pseudo-second-order kinetic model in which the chemisorption mechanism is significantly influenced by electrostatic interactions. With a maximum sorption capacity of 681.3 mg/g after being annealed at 250 °C, these composites demonstrated the efficacy of GR-based nanocomposites for treating groundwater [44].

Furthermore, the immobilization and removal of As(III) onto A-doped GR (AG: AC93H24) nano-adsorbents, containing first-row transition metals (A: Sc–Zn), were described by theoretical calculations in both gas phase and explicit/implicit solvent media. By primarily using bi-dentate surface complexation similar to that found on mineral surfaces, the AG adsorbents demonstrated a strong affinity for As(III) with adsorption energies ranging from 1.2 to 2.0 eV in the gas phase. In addition, compared to intrinsic GR, the adsorption stability rose by at least 229% [45]. Through Direct Current (DC) and Alternating Current (AC) electro-deposition, an electrochemical technique employing a GO-modified carbon felt (CF/GO) electrode has been developed to address heavy metal contamination at low and high concentrations. The high density of surface functional groups provided by GO results in an adsorption capacity that is two orders of magnitude higher than that of conventional adsorption methods (29 g of heavy metal per 1 g of GO). Even in situations where there is minimal heavy metal contamination at the point of use, this technique successfully lowers pollution from single HMIs (Cu, Pb, and Cd) as well as multiple-ion combinations to levels deemed safe for drinking water by using a CF/GO anode for dc electro-deposition. Compared to traditional adsorption techniques, this technology can handle a far wider variety of heavy metal contaminants in point-of-use water. Over 99.9% of HMIs may be recovered by dc electro-deposition from highly polluted industrial wastewater, as shown in Fig. 1. Additionally, the electro-deposition approach can recover Cu, Pb, and Cd preferentially by varying the voltage and ac frequency, increasing the heavy metal removal process's value and efficiency [46].

2.3 Graphene as a Candidate for Membrane Filtration Technique for Purification of Water

Numerous studies in the literature have shown how to create monolayer GR membranes for desalination and ultra-filtration (UF). Celebi et al. [47], for instance, created holes in a nanoporous GR membrane by focusing an ion beam. With a hole size of 50 nm and a porosity of 4.7%, this membrane produced a permeate flux of roughly 10.1 m^3/m^2 h bar. Compared to what has been documented for cellulose, acrylic, and polysulfone ultra-filtration membranes, this performance is five to seven times better. In another study, researchers employed oxygen plasma etching to create sub-nanometer-sized pores (1 nm) in monolayer GR. At 40 °C, this nanoporous GR membrane totally rejected sodium, lithium, potassium, and chloride ions while achieving an amazing permeate flow of up to 3600 m^3/m^2 h [48]. Interestingly, GR-based materials have been modified with NMs, emerging as a viable alternative to enhance membrane permeability. These NMs can be inorganic NPs or organic compounds [49]. According to recent research, GR may find application in the creation of highly permeable and selective separation membranes. These membranes could outperform inorganic filters and polymer-based membranes in water purification. By reducing the thickness of primary GR-based membrane to approximately 50 nm, membranes with high permeation flux have been achieved. Because of their improved surface characteristics and customizable functionalities, low-dimensional GR NMs allow for membrane modification that strikes a compromise between selectivity and permeability. GR-based frameworks with polar functional groups, which are regarded as an essential component of nanotechnology, enable the development of a perfect membrane [50]. Furthermore, the integration of NPs creates frictionless capillary channels that can improve the membranes' ability to infuse substances rapidly. Additionally, due to its distinct lattice structure, GR has developed into an intriguing functional membrane substance for water filtration [51]. A wide variety of derivatives of GR have also been researched in order to enhance the capabilities of the material even more. The characteristics of GR can be customized through functionalization and doping for a number of high performance uses, including membrane technology. One of the key GR derivatives that largely preserve the properties of

GR is GO. GO is water dispersible as it has polar functional groups with oxygen in them. It is possible to tailor the amount of oxidation in GO during preparation, that is essential for a subsequent modification of derivatives of GR [52]. In addition, GO derivatives exhibit remarkable membrane characteristics such anti-fouling, high Cl_2 tolerance, and antibacterial activity, extending the lifetime of the membrane [53]. Fouling-resistant membranes with adjustable surface functions have the potential to totally eliminate the pre-treatment process, which would drastically lower the energy usage of desalination plants [54]. For the treatment of brackish water using a nano-filtration membrane, GR membranes have currently demonstrated enormous potential [55]. The efficiency of GR membranes for Reverse Osmosis (RO) desalination has been shown by theoretical research and molecular simulations [56]. The structure of GO contains both sp^3 and sp^2, i.e., aliphatic as well as aromatic carbon atoms, along with OFGs. This composition allows for chemical reactions and surface modifications, enabling the creation of various assembly morphologies tailored for specific separation applications. [57]. Functional groups with polarity enhance the dispersion and hydrophilicity of GO in aqueous media. GO is not only cost-effective but also has the potential for large-scale production, making it an attractive material for research. By controlling the synthesis of GO through variations in oxidation degree, size, and functional groups it is possible to analyze and comprehend the separation mechanisms of lamellar structure, resulting in high separation efficiency [58]. For high-capacity water treatment applications, GO membranes that were chemically reduced with theanine and cross-linked with tannic acid were created. The remarkable water permeability of approximately 10,000 L/m^2 h bar demonstrated by these membranes is 10–1000 times greater than any previously documented values for GO-based materials used in commercial applications. With a 100% rejection rate for the tested dyes, such as MB and rhodamine B, they also demonstrated remarkable separation capabilities [59].

3 Graphene-Based Catalysts for Water Pollutant Removal

Because of their narrower energy gaps and lower electron–hole pair recombination rates, rGO is used in advanced oxidation processes to accelerate the photodegradation of synthetic organic pollutants, such as different compounds and colors. Increased photo-degradation and better light absorption are the outcomes of altering the valence and conduction band levels [60]. In a similar application, rGO derived from Murraya koenigii leaf extracts (MkrGO) was employed to remove MB and Methylene Orange (MO) dyes. The synthetic MkrGO successfully degraded up to 80% of MO dye and 77% of MB dye. Sunlight exposure initiated degradation, with the most significant breakdown occurring within the first 20–30 min and continuing over a two-hour period. To mitigate membrane fouling and eliminate organic contaminants, photocatalysis has been integrated with membrane separation techniques. This combination addresses the challenge of photocatalyst recovery in photocatalysis technology while also tackling membrane fouling, a key issue that can be alleviated by developing nano-filtration membranes with photocatalytic degradation capabilities [61]. The TiO_2-coated GO membrane (GOT/P25) [62], the O-g-C_3N_4/GO/N-CNT membrane [63], the 1D graphitic carbon nitride nanotube (g-C_3N_4-NT) intercalated rGO membrane [64], and the GO/palygorskite/g-C_3N_4@$Bi_2O_2CO_3$ hetero-structured membrane [65] are a few achievements that focus on the simple coupling of rGO or GO nanosheets with photocatalysts. Using a mixed-dimensional assembly technique, a new GR-based membrane was created, utilizing AgNPs linked to g-C_3N_4 as photocatalysts and pillars. The water transport channels in the rGO and g-C_3N_4 laminates are greatly improved by the even dispersion of AgNPs on g-C_3N_4. The resultant rGO/AgNP@g-C_3N_4 nano-filtration membrane has remarkable water permeability, photocatalytic self-cleaning capabilities, and remarkable flexibility and structural stability. The rGO/AgNP@g-C_3N_4 membrane has a water permeability of 210.9 Lm^{-2} h^{-1} bar^{-1}, which is much higher than that of the pure rGO membrane, which has a permeability of 78.4 Lm^{-2} h^{-1} bar^{-1}. This is due to the rough surface and larger interlayer spacing. After only one hour of exposure to visible light, the membrane may recover from flux loss brought on by pollutant adsorption, attaining an exceptional Flux Recovery Ratio (FRR) of 98.1%. Furthermore, in a cross-flow photocatalytic nano-filtration device, the high flux of the rGO/AgNP@g-C_3N_4 membrane does not change. Its total performance is improved by the efficient cooperation of membrane separation, photocatalytic regeneration, and pollutant degradation by OH and superoxide (O_2^-) radicals [61].

3.1 Factors Influencing the Activity of Graphene-Based Nanomaterials

The properties of the background solution are just as crucial to wastewater treatment as those of organic pollutants and GR-based compounds. Temperature, pH, ionic strength, and the presence of natural organic matter are important variables that affect adsorption [66]. Changes in the surrounding pH have a considerable effect on the adsorption of inorganic pollutants. Therefore, wastewater treatment systems must be carefully assessed, as different chemicals respond uniquely to varying pH levels. Generally, cationic contaminants are more effective at higher pH values, while anionic pollutants are better adsorbed at lower pH levels. Temperature also plays a significant role in any physical or chemical process involved in adsorption. For instance, Elsagh et al. used GR, GO, and carboxylated single-walled CNTs (SWCNT–COOH) in comparative absorption tests to extract Basic Red 46 (BR46) from contaminated water bodies [67]. They found that SWCNT–COOH and GO serve as highly effective adsorbents for BR46. The study found that although the maximal adsorption capacity does not vary as the temperature rises, the adsorption process does. The adsorption mechanism was found to be physisorption, using a pseudo first-order kinetics model. Additionally, the surrounding is also a critical factor in determining the efficiency of any process. Ions present in the aqueous solution can compete for adsorption sites, which may negatively impact overall performance of the catalyst or adsorbent. For example, sorption capacity for As(III) was observed to decrease by 17% in the presence of coexisting anions, such as F^-, when using the RGO/$Mg(OH)_2$–MgO nanocomposite [44].

GR is several tens or hundreds of times more expensive than MWCNTs, whether it is manufactured using Hummer's approach or the CVD procedure [68]. However, compared to the previously stated MWCNTs, AC-based adsorbents are significantly less expensive. Therefore, GR-based adsorbents could only be used realistically or economically if their lifetime was hundreds of times longer than that of AC-based adsorbents [69] or if their operating cost was significantly cheaper. Fortunately, GR-based adsorbents require significantly lower energy and heat costs for desorption of organic compounds in comparison to AC-based adsorbents, making their practical application feasible. Cost is a crucial factor in determining commercial viability, with the total expense of an adsorption unit comprising the one-time costs of the adsorbents, apparatus installation, operation, and consumables like steam, electricity, water, and nitrogen. Notably, the adsorbent costs can account for one-third to one-half of the initial investment. Commercial GR-based adsorbents are typically created by combining GR powder with inorganic oxides and undergoing drying and calcination. However, these powders are often produced through Chemical Vapor Deposition (CVD) on a support, usually an inorganic oxide, which is then treated with acids or bases for removal during post-purification. This approach can lead to overlapping operational phases. By integrating these processes, manufacturers can achieve cost and energy savings while enhancing control over GR quality, minimizing issues such as layer stacking and poor dispersion.

4 Pros and Cons of Graphene-Based Nanomaterials

The cost of production, effectiveness, and environmental impact are some of the considerations when using GR as a carbon-based nanomaterial for contaminant removal. Because of its cheaper manufacturing costs, GO demonstrates environmental improvements similar to those of pure GR. GR is a better material than CNTs for wastewater treatment. GR-based materials have an edge over CNTs as far as adsorption is concerned, as the innermost walls of the CNTs are inaccessible to adsorbents, while the single layer of GR has two basal planes that can be used for impurity adsorption [70].

5 Conclusions and Future Perspective

On one hand, the growing population and industrialization have made water pollution a pressing concern today. On the other hand, the demand for pure and clean water is rising sharply. Therefore, wastewater purification and treatment have become crucial to meet this increasing demand and to mitigate environmental impacts. Graphene-based nanomaterials hold significant potential in this area by adsorbing, degrading, and filtering pollutants. In addition to investigating the factors that affect these nanomaterials' efficacy in aquatic environments, numerous studies have investigated the usage of these materials for the removal of various organic and inorganic pollutants using techniques like adsorption, photocatalytic degradation, and membrane filtration. Due to their impressive performance, properties, and availability, these materials could represent the future of wastewater treatment technologies. However, challenges such as their cost and cytotoxicity must be addressed.

References

1. Bhat SA, Sher F, Hameed M, Bashir O, Kumar R, Vo DV, Ahmad P, Lima EC (2022) Sustainable nanotechnology based wastewater treatment strategies: achievements, challenges and future perspectives. Chemosphere 1(288):132606
2. Lai Y, Wang F, Zhang Y, Ou P, Wu P, Fang Q, Chen Z, Li S (2019) UiO-66 derived N-doped carbon NPs coated by PANI for simultaneous adsorption and reduction of hexavalent chromium from waste water. Chem Eng J 15(378):122069
3. Oki T, Kanae S (2006) Global hydrological cycles and world water resources. Science 313(5790):1068–1072
4. Bhat SA, Pandit BA, Khan JN, Kumar R, Jan R (2017) Water requirements and irrigation scheduling of maize crop using CROPWAT model. Int J Curr Microbiol Appl Sci 6(11):1662–1670
5. Elbeltagi A, Azad N, Arshad A, Mohammed S, Mokhtar A, Pande C, Etedali HR, Bhat SA, Islam AR, Deng J (2021) Applications of Gaussian process regression for predicting blue water footprint: case study in Ad Daqahliyah, Egypt. Agric Water Manag 1(255):107052
6. Edokpayi JN, Makungo R, Mathivha F, Rivers N, Volenzo T, Odiyo JO (2020) Influence of global climate change on water resources in South Africa: toward an adaptive management approach. In: Water conservation and wastewater treatment in BRICS Nations. Elsevier, pp 83–115
7. Yadav A, Chowdhary P, Kaithwas G, Bharagava RN (2017) Toxic metals in the environment: threats on ecosystem and bioremediation approaches. In: Handbook of metal-microbe interactions and bioremediation. CRC Press, pp 128–141
8. Acharya K, Blackburn A, Mohammed J, Haile AT, Hiruy AM, Werner D (2020) Metagenomic water quality monitoring with a portable laboratory. Water Res 1(184):116112
9. Basheer AA (2018) New generation nano-adsorbents for the removal of emerging contaminants in water. J Mol Liq 1(261):583–593
10. Shah AI, Dar MU, Bhat RA, Singh JP, Singh K, Bhat SA (2020) Prospectives and challenges of wastewater treatment technologies to combat contaminants of emerging concerns. Ecol Eng 1(152):105882
11. Liu F, Zhu S, Li D, Chen G, Ho SH (2020) Detecting ferric iron by microalgal residue-derived fluorescent nanosensor with an advanced kinetic model. Iscience 3(6)

12. Kadam A, Saratale RG, Shinde S, Yang J, Hwang K, Mistry B, Saratale GD, Lone S, Kim DY, Sung JS, Ghodake G (2019) Adsorptive remediation of cobalt oxide NPs by magnetized α-cellulose fibers from waste paper biomass. Biores Technol 1(273):386–393
13. Kalaitzidou K, Zouboulis A, Mitrakas M (2020) Cost evaluation for Se (IV) removal, by applying common drinking water treatment processes: Coagulation/precipitation or adsorption. J Environ Chem Eng 8(5):104209
14. Lam E, Luong JH (2014) Carbon materials as catalyst supports and catalysts in the transformation of biomass to fuels and chemicals. ACS Catal 4(10):3393–3410
15. Ma Y, Han J, Wang M, Chen X, Jia S (2018) Electrophoretic deposition of graphene-based materials: a review of materials and their applications. J Mater 4(2):108–120
16. Nupearachchi CN, Mahatantila K, Vithanage M (2017) Application of graphene for decontamination of water; implications for sorptive removal. Groundw Sustain Dev 1(5):206–215
17. Li MF, Liu YG, Zeng GM, Liu N, Liu SB (2019) Graphene and graphene-based nanocomposites used for antibiotics removal in water treatment: a review. Chemosphere 1(226):360–380
18. Puri N, Mishra SK, Niazi A, Srivastava AK, Rajesh (2015) Physicochemical characteristics of reduced graphene oxide based Pt-NPs-conducting polymer nanocomposite film for immunosensor applications. J Chem Technol Biotechnol 90(9):1699–1706
19. Pendolino F, Armata N (2017) Graphene oxide in environmental remediation process. Springer, Switzerland
20. Jannesari M, Akhavan O, Hosseini HR (2018) Graphene oxide in generation of nanobubbles using controllable microvortices of jet flows. Carbon 1(138):8–17
21. Vieira Y, Ceretta MB, Foletto EL, Wolski EA, Silvestri S (2020) Application of a novel rGO–$CuFeS_2$ composite catalyst conjugated to microwave irradiation for ultra-fast real textile wastewater treatment. J Water Process Eng 1(36):101397
22. Rashed MN (2013) Adsorption technique for the removal of organic pollutants from water and wastewater. Org Pollut-Monit Risk Treat 30(7):167–194
23. Kumar S, Kaushik RD, Purohit LP (2022) ZnO–CdO nanocomposites incorporated with graphene oxide nanosheets for efficient photocatalytic degradation of bisphenol A, thymol blue and ciprofloxacin. J Hazard Mater 15(424):127332
24. Luo Z, Yao B, Yang X, Wang L, Xu Z, Yan X, Tian L, Zhou H, Zhou Y (2022) Novel insights into the adsorption of organic contaminants by biochar: a review. Chemosphere 1(287):132113
25. Ng LY, Chua HS, Ng CY (2021) Incorporation of graphene oxide-based nanocomposite in the polymeric membrane for water and wastewater treatment: a review on recent development. J Environ Chem Eng 9(5):105994
26. Sadegh H (2017) Development of graphene oxide from graphite: a review on synthesis, characterization and its application in wastewater treatment. Rev Adv Mater Sci 49(1):38–43
27. Jiang LH, Liu YG, Zeng GM, Xiao FY, Hu XJ, Hu X, Wang H, Li TT, Zhou L, Tan XF (2016) Removal of 17β-estradiol by few-layered graphene oxide nanosheets from aqueous solutions: external influence and adsorption mechanism. Chem Eng J 15(284):93–102
28. Khan S, AchazhiyathEdathil A, Banat F (2019) Sustainable synthesis of raphene-based adsorbent using date syrup. Sci Rep 9(1):18106
29. Queiroz RN, Prediger P, Vieira MG (2022) Adsorption of polycyclic aromatic hydrocarbons from wastewater using graphene-based nanomaterials synthesized by conventional chemistry and green synthesis: a critical review. J Hazard Mater 15(422):126904
30. Martínez-Álvarez I, Le Menach K, Devier MH, Barbarin I, Tomovska R, Cajaraville MP, Budzinski H, Orbea A (2021) Uptake and effects of graphene oxide nanomaterials alone and in combination with polycyclic aromatic hydrocarbons in zebrafish. Sci Total Environ 25(775):145669
31. Song T, Tian W, Qiao K, Zhao J, Chu M, Du Z, Wang L, Xie W (2021) Adsorption behaviors of polycyclic aromatic hydrocarbons and oxygen derivatives in wastewater on N-doped reduced graphene oxide. Sep Purif Technol 1(254):117565
32. Chen H, Liu T, Meng Y, Cheng Y, Lu J, Wang H (2020) Novel graphene oxide/aminated lignin aerogels for enhanced adsorption of malachite green in wastewater. Colloids Surf A: PhysChemical Eng Asp 20(603):125281

33. Bu J, Yuan L, Zhang N, Liu D, Meng Y, Peng X (2020) High-efficiency adsorption of methylene blue dye from wastewater by a thiosemicarbazide functionalized graphene oxide composite. Diam Relat Mater 1(101):107604
34. Zheng AL, Phromsatit T, Boonyuen S, Andou Y (2020) Synthesis of silver NPs/porphyrin/reduced graphene oxide hydrogel as dye adsorbent for wastewater treatment. FlatChem 1(23):100174
35. Sirajudheen P, Karthikeyan P, Ramkumar K, Meenakshi S (2020) Effective removal of organic pollutants by adsorption onto chitosan supported graphene oxide-hydroxyapatite composite: a novel reusable adsorbent. J Mol Liq 15(318):114200
36. Gupta RK, Alqahtani FH, Dawood OM, Carini M, Criado A, Prato M, Garlapati SK, Jones G, Sexton J, Persaud KC, Dang C (2020) Suspended graphene arrays for gas sensing applications. 2D Materials 8(2):025006
37. Mallikarjuna K, Reddy LV, Al-Rasheed S, Mohammed A, Gedi S, Kim WK (2021) Green synthesis of reduced graphene oxide-supported palladium NPs by coleus amboinicus and its enhanced catalytic efficiency and antibacterial activity. Crystals 11:134
38. Briffa J, Sinagra E, Blundell R (2020) Heavy metal pollution in the environment and their toxicological effects on humans. Heliyon 6(9)
39. Manikandan VS, Boateng E, Durairaj S, Chen A (2022) Electrochemical sensing of vanillin based on fluorine-doped reduced graphene oxide decorated with gold NPs. Foods 11(10):1448
40. Xu X, Lv X, Tan F, Li Y, Geng C, Cui B, Fang Y (2021) Simultaneous electrochemical detection of multiple heavy-metal ions based on fur-fural/reduced graphene oxide composites. J Electrochem Soc 168(12):126516
41. Bloor JM, Handy RD, Awan SA, Jenkins DF (2021) Graphene oxide biopolymer aerogels for the removal of lead from drinking water using a novel nano-enhanced ion exchange cascade. Ecotoxicol Environ Saf 15(208):111422
42. Wang Y, Ding G, Lin K, Liu Y, Deng X, Li Q (2021) Facile one-pot synthesis of ultrathin carbon layer encapsulated magnetite nanoparticle and graphene oxide nanocomposite for efficient removal of metal ions. Sep Purif Technol 1(266):118550
43. Priyadharshini SD, Manikandan S, Kiruthiga R, Rednam U, Babu PS, Subbaiya R, Karmegam N, Kim W, Govarthanan M (2022) Graphene oxide-based nanomaterials for the treatment of pollutants in the aquatic environment: recent trends and perspectives–a review. Environ Pollut 1(306):119377
44. Puri N, Gupta A, Mishra A (2021) Recent advances on nano-adsorbents and nanomembranes for the remediation of water. J Clean Prod 1(322):129051
45. Cortes-Arriagada D, Mella A (2019) Performance of doped graphene nanoadsorbents with first-row transition metals (ScZn) for the adsorption of water-soluble trivalent arsenicals: a DFT study. J Mol Liq 15(294):111665
46. Wan K, Wang Y, Liu C, Wei C, Lv S, Tang X, Fang T, Zhao J, Wei G, Qi P, Liu X (2022) Facile synthesis of hierarchical Ti3C2@ FeOOH nanocomposites for antimony contaminated wastewater treatment: performance, mechanisms, reutilization, and sustainability. Chem Eng J 15(450):138038
47. Celebi K, Buchheim J, Wyss RM, Droudian A, Gasser P, Shorubalko I, Kye JI, Lee C, Park HG (2014) Ultimate permeation across atomically thin porous graphene. Science 344(6181):289–292
48. Asif MB, Iftekhar S, Maqbool T, Pramanik BK, Tabraiz S, Sillanpää M, Zhang Z (2022) Two-dimensional nanoporous and lamellar membranes for water purification: reality or a myth? Chem Eng J 15(432):134335
49. Athanasekou CP, Morales-Torres S, Likodimos V, Romanos GE, Pastrana-Martinez LM, Falaras P, Dionysiou DD, Faria JL, Figueiredo JL, Silva AM (2014) Prototype composite membranes of partially reduced graphene oxide/TiO_2 for photocatalytic ultrafiltration water treatment under visible light. Appl Catal B 1(158):361–372
50. Zhu X, Yang K, Chen B (2017) Membranes prepared from graphene-based nanomaterials for sustainable applications: a review. Environ Sci Nano 4(12):2267–2285

51. Sivakumar M, Yadav S, Hung WS, Lai JY (2020) One-pot eco-friendly synthesis of edge-carboxylate graphene via dry ball milling for enhanced removal of acid and basic dyes from single or mixed aqueous solution. J Clean Prod 1(263):121498
52. Park MJ, Nisola GM, Seo DH, Wang C, Phuntsho S, Choo Y, Chung WJ, Shon HK (2021) Chemically cross-linked graphene oxide as a selective layer on electrospun polyvinyl alcohol nanofiber membrane for nanofiltration application. Nanomaterials 11(11):2867
53. Chen X, Qiu M, Ding H, Fu K, Fan Y (2016) A reduced graphene oxide nanofiltration membrane intercalated by well-dispersed carbon nanotubes for drinking water purification. Nanoscale 8(10):5696–5705
54. Nicolaï A, Sumpter BG, Meunier V (2014) Tunable water desalination across graphene oxide framework membranes. Phys Chem Chem Phys 16(18):8646–8654
55. Yadav S, Ibrar I, Altaee A, Samal AK, Ghobadi R, Zhou J (2020) Feasibility of brackish water and landfill leachate treatment by GO/MoS2-PVA composite membranes. Sci Total Environ 25(745):141088
56. Cohen-Tanugi D, Grossman JC (2015) Nanoporous graphene as a reverse osmosis membrane: recent insights from theory and simulation. Desalination 15(366):59–70
57. Xu JC, Ma Q, Chen C, Wu QT, Long XX (2020) Cadmium adsorption behavior of porous and reduced graphene oxide and its potential for promoting cadmium migration during soil electrokinetic remediation. Chemosphere 1(259):127441
58. Wei Y, Zhang Y, Gao X, Ma Z, Wang X, Gao C (2018) Multilayered graphene oxide membranes for water treatment: a review. Carbon 1(139):964–981
59. Thebo KH, Qian X, Zhang Q, Chen L, Cheng HM, Ren W (2018) Highly stable graphene-oxide-based membranes with superior permeability. Nat Commun 9(1):1486
60. Khan SA, Yu Z, Belhadi A, Mardani A (2020) Investigating the effects of renewable energy on international trade and environmental quality. J Environ Manage 15(272):111089
61. Chen C, Chen L, Zhu X, Chen B (2022) Graphene nanofiltration membrane intercalated with AgNP@ g-C3N4 for efficient water purification and photocatalytic self-cleaning performance. Chem Eng J 1(441):136089
62. Xu C, Xu Y, Zhu J (2014) Photocatalytic antifouling graphene oxide-mediated hierarchical filtration membranes with potential applications on water purification. ACS Appl Mater Interfaces 6(18):16117–16123
63. Qu L, Zhu G, Ji J, Yadav TP, Chen Y, Yang G, Xu H, Li H (2018) Recyclable visible light-driven Og-C3N4/graphene oxide/N-carbon nanotube membrane for efficient removal of organic pollutants. ACS Appl Mater Interfaces 10(49):42427–42435
64. Wei Y, Zhu Y, Jiang Y (2019) Photocatalytic self-cleaning carbon nitride nanotube intercalated reduced graphene oxide membranes for enhanced water purification. Chem Eng J 15(356):915–925
65. Cai Y, Chen D, Li N, Xu Q, Li H, He J, Lu J (2020) A self-cleaning heterostructured membrane for efficient oil-in-water emulsion separation with stable flux. Adv Mater 32(25):2001265
66. Cao M, Shen Y, Yan Z, Wei Q, Jiao T, Shen Y, Han Y, Wang Y, Wang S, Xia Y, Yue T (2021) Extraction-like removal of organic dyes from polluted water by the graphene oxide/PNIPAM composite system. Chem Eng J 1(405):126647
67. de Figueiredo NT, Dalarme NB, da Silva PM, Landers R, Picone CS, Prediger P (2020) Novel magnetic chitosan/quaternary ammonium salt graphene oxide composite applied to dye removal. J Environ Chem Eng 8(4):103820
68. Tai Z, Yang J, Qi Y, Yan X, Xue Q (2013) Synthesis of a graphene oxide–polyacrylic acid nanocomposite hydrogel and its swelling and electroresponsive properties. RSC Adv 3(31):12751–12757
69. Yin Z, Cui C, Chen H, Duoni, Yu X, Qian W (2020) The application of carbon nanotube/graphene-based nanomaterials in wastewater treatment. Small 16(15):1902301
70. Sitko R, Zawisza B, Malicka E (2013) Graphene as a new sorbent in analytical chemistry. TrAC Trends Anal Chem 1(51):33–43

Composite Nanoclay Materials for Removal of Ions from Water

A. H. Bhat, Amal S. Al Rahbi, Veena Doss, Bhupinder Kaur, Showkat Ahmad Bhawani, and Geetha Devi

Abstract Nanoclay, a material derived from natural clays, exhibits strong potential for wastewater treatment due to its exceptionally high surface area, making it highly efficient at removing toxic metal ions, microbes, and dissolved pollutants. This chapter reviews recent progress in nanoclay-based adsorbents, exploring their applications and the mechanisms behind ion removal. The adsorption process, governed by charge chemistry, involves multiple pathways, including physical adsorption, micro-precipitation, and chemisorption. Additionally, the chapter evaluates key factors that influence adsorption performance, such as nano-adsorbent surface properties and external chemical conditions. By providing these insights, the discussion aims to advance the understanding of nanoclay composites and their role in water purification.

Keywords Nanoclay · Pollutants · Composites · Adsorption

A. H. Bhat (✉) · A. S. Al Rahbi · V. Doss
College of Applied Science and Pharmacy, University of Technology and Applied Sciences (UTAS) Applied Science Department, Muscat, Sultanate of Oman
e-mail: aamir.bhat@utas.edu.om

A. S. Al Rahbi
e-mail: amal.alrahbi@utas.edu.om

V. Doss
e-mail: veena.doss@utas.edu.om

B. Kaur
Faculty of Medicine Quest International University Perak, Ipoh, Malaysia
e-mail: bhupinder.kaur@qui.edu.my

S. A. Bhawani
Faculty of Resource Science and Technology, Universiti Malaysia Sarawak, Kota Samarahan, Malaysia
e-mail: sabhawani@gmail.com

G. Devi
College of Engineering, University of Science and Technology, Muscat, Oman
e-mail: geethadrvi@nu.edu.om

S. A. Bhawani and M. Jawaid (eds.), *Nanomaterials for Separation of Hazardous Contaminants from Wastewater*, Smart Nanomaterials Technology,
https://doi.org/10.1007/978-981-95-6079-0_3

1 Introduction

The release of toxic heavy metal ions into aquatic ecosystems from industrial sectors such as textiles, fertilizers, mining, refining, and printing presents a profound threat to environmental integrity, ecological stability, and public health. Owing to their non-biodegradable and bioaccumulative nature, these metals persist in environmental matrices over extended periods, thereby exacerbating their toxicological impact [1–3]. Mercury, among the most hazardous of these contaminants, has a stringent permissible limit of only 2.0 parts per billion (ppb) in potable water. Chronic exposure to mercury is associated with severe neurotoxicity and deleterious effects on pulmonary, renal, and other organ systems. Although various remediation techniques—such as ion exchange, membrane filtration, and reverse osmosis—are employed for heavy metal removal, adsorption continues to be the preferred method due to its operational simplicity, economic viability, and high efficiency [4–9].

In recent years, inorganic–organic hybrid nanocomposites (NCs) integrating clay-based materials have emerged as a promising class of adsorbents in the domain of environmental remediation. These materials offer significant advantages, including natural abundance, mechanical robustness, thermal stability, and structural versatility. Nanoclays, derived from finely dispersed mineral phases within natural clays, are primarily composed of layered silicate structures. These layers, formed by corner-sharing silicon-oxygen tetrahedra [5], confer high surface area and exceptional adsorptive properties, enabling effective sequestration of trace-level contaminants. The unique physicochemical characteristics of clay minerals render them highly applicable in nanotechnology, environmental engineering, and a wide array of industrial processes [6].

Clay minerals have historically played a crucial role across diverse scientific and technological fields due to their capacity to enhance process efficiency, improve material performance, and reduce production costs—thereby aligning with sustainability objectives. Nanoclays, a prominent subclass of clay minerals, are extensively distributed in nature and have been utilized since antiquity, notably in prehistoric ceramics and in the production of goods using kaolin (China clay) as early as the third century in China. Structurally, individual clay layers comprise two to four sheets of silica tetrahedra or aluminum hydroxide octahedra. These hydrous silicates exhibit fine, plate-like morphologies arranged in characteristic sheet-like (phyllosilicate) structures.

These phyllosilicate structures are analogous to the pages of a book, held in place by Van der Waals forces that maintain consistent interlayer spacing [8]. The interlayers feature ionic substitutions within the crystal lattice, generating negatively charged surfaces counterbalanced by exchangeable cations situated in the inter-lamellar regions. These cations—commonly including sodium (Na^+), potassium (K^+), magnesium (Mg^{2+}), and calcium (Ca^{2+})—can be readily replaced through ion-exchange mechanisms, facilitating surface functionalization for specific environmental applications. While material performance is generally governed by chemical composition, in clay-based systems, physical properties such as particle size, morphology, and layer structure are critical for defining their functional capabilities. Various separation and purification techniques, including high-energy stirring, centrifugation, freeze-drying, and ultracentrifugation, are routinely employed for the isolation and refinement of clay minerals [9].

2 Classification and Properties of Clay

Clay minerals exhibit a diverse array of physical characteristics, including shrinkage upon drying, plasticity, hardness, porosity, and notable adsorption capacity [10]. Their nanoscale particle size, high surface area, and intrinsic porosity facilitate extensive physicochemical interactions at the surface level. These interactions are primarily governed by strong electrostatic repulsion and selective cation exchange mechanisms, both of which play a critical role in the adsorption of pollutants. The pronounced porosity of nanoclay significantly enhances its affinity for binding with various contaminants. Structurally, clay minerals consist of hydrous aluminosilicates, incorporating various metal and mineral oxides, which contribute to their inherent stability and reactivity. Based on these structural and compositional features, clay minerals are categorized into distinct groups—most notably kaolinite, montmorillonite, and illite—according to Grim's Classification system [11]. Each of these groups is defined by its characteristic layer structure and compositional attributes, which determine its specific functional behavior.

2.1 Structures of Nanoclays

The structural configuration of a mineral is governed by its chemical composition, the spatial arrangement of constituent molecules and ions, and the electrostatic forces that maintain their cohesion [12]. Approximately 99% of the Earth's crust is composed of eight predominant elements: oxygen (O^{2-}), silicon (Si^{4+}), aluminum (Al^{3+}), iron (Fe^{2+}), magnesium (Mg^{2+}), calcium (Ca^{2+}), sodium (Na^{+}), and potassium (K^{+}). Clay minerals are primarily constructed from sheets of silicon and aluminum oxides (Al_2O_3), classifying them within the phyllosilicate group [13]. Their subclassification is determined by the specific arrangement of silicate layers and the integration of metal oxide and hydroxyl ion sheets [14]. The basic structural unit of most clay minerals consists of a 1:1 layer configuration, comprising one tetrahedral silicate sheet bonded to one octahedral sheet [15]. In the octahedral sheet, a central plane of aluminum or magnesium ions is sandwiched between two planes of hydroxyl ions (OH^{-}), forming a hexagonally coordinated arrangement [16]. A 2:1 layer structure, which includes two tetrahedral sheets flanking an octahedral sheet, characterizes mineral groups such as smectite and mica—both of which hold significant relevance within the broader clay mineral family.

2.2 Clay Minerals: Structural Properties and Applications

The primary composition of clay minerals typically involves two main geometric structures: tetrahedral and octahedral sheets, which are arranged in various configurations. Variations in these structural arrangements give rise to different types of clay minerals. Below, we discuss some of the major groups of clay minerals.

Kaolin

The kaolinite group comprises a diverse array of clay minerals, including dickite, nacrite, kaolinite, and halloysite, all of which originate from the weathering and decomposition of orthoclase feldspar. Among these, dickite and nacrite are typically formed through hydrothermal processes and are frequently associated with kaolinite deposits [17]. These minerals share a common structural framework composed of alternating silica (tetrahedral) and alumina (octahedral) sheets, arranged in a 1:1 layer configuration. This arrangement categorizes kaolinite as a 1:1 type clay mineral, as illustrated in Fig. 1.

Kaolin, a widely utilized member of the kaolinite group—also known as China clay—holds significant industrial relevance. Its utility as a filler material in the paper and plastic industries is attributed to its favorable particle size and morphology, which lend versatility across multiple applications. The mineral's chemical stability, fine particle size, and high whiteness make it especially suitable for use in ceramics, catalysts, adhesives, and water purification processes, particularly as an adsorbent. The general chemical formula of kaolinite is $Al_2Si_2O_5(OH)_4$, comprising the major constituents SiO_2, Al_2O_3, and H_2O, with minor impurities such as Mg, K, Fe, Ti, and others.

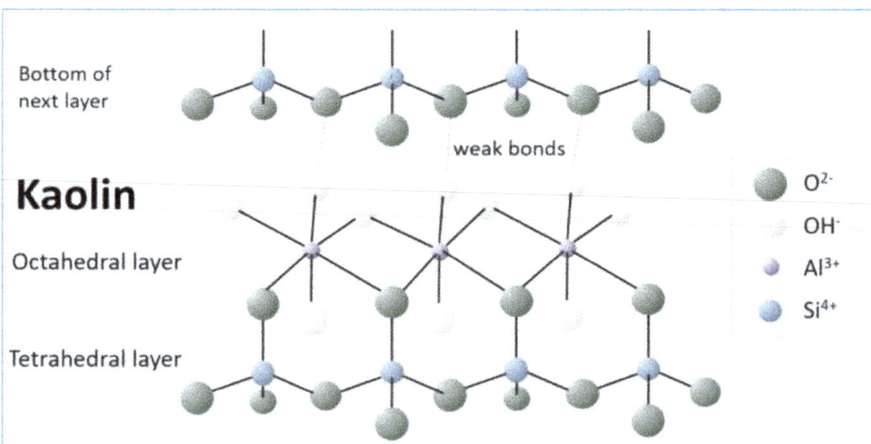

Fig. 1 Kaolin clay structure

Smectite and Montmorillonite

Smectite constitutes the largest and most significant class of phyllosilicate clay minerals, distinguished from the kaolinite group by its unique chemical composition and structural properties. Smectite minerals exhibit morphologies ranging from equidimensional to elongated forms and encompass several compositional variants, including iron-rich montmorillonite (nontronite), magnesium-rich montmorillonite (saponite), and lithium-rich montmorillonite (hectorite). A defining characteristic of this group is the occurrence of isomorphous substitutions within both octahedral and tetrahedral sheets: Magnesium and iron frequently substitute for aluminum in the octahedral layers, while aluminum and silica substitutions occur within the tetrahedral framework.

The principal members of the group—sodium montmorillonite and calcium montmorillonite—exhibit subtle but functionally significant structural differences. Sodium montmorillonite typically incorporates a single layer of interlayer water, whereas calcium montmorillonite accommodates two layers. The presence of a double water layer in calcium montmorillonite results in reduced swelling capacity and lower viscosity compared to its sodium analogue.

Owing to their structural variability and physicochemical versatility, smectite minerals are employed extensively across environmental and industrial sectors. Applications include their use as high-performance adsorbents, organoclays (nanoclays) for wastewater remediation, functional additives in food processing, moisture control agents (desiccants), and stabilizers in emulsion systems. The general structural formula for the smectite group is expressed as $(Ca,Na,H)(Al,Mg,Fe,Zn)_2(Si,Al)_4O_{10}(OH)_2{\cdot}XH_2O$ (Fig. 2).

Chlorite

The chlorite group is defined by a 2:1 layer silicate structure, comprising alternating tetrahedral and octahedral sheets, and is generally represented by the formula $(Mg,Fe,Li)_6AlSi_3O_{10}(OH)_8$. Individual minerals within this group exhibit distinct chemical compositions, including chamosite $(Fe^{2+},Mg)_5Al(AlSi_3O_{10})(OH)_8$, cookeite $LiAl_5Si_3O_{10}(OH)_8$, orthochamosite $(Fe^{2+},Mg,Fe^{3+})_5Al(AlSi_3O_{10})(OH,O)_8$, and clinochlore $(Mg,Fe^{2+})_5Al_2Si_3O_{10}(OH)_8$. Chlorite minerals occur predominantly within schistose metamorphic rocks and encompass a range of clay mineral species, including clinochlore, cookeite, donbassite, franklinfurnaceite, nimite, orthochamosite, sudoite, baileychlore, and borocookeite.

Palygorskite and sepiolite, although also classified as 2:1 layer silicates, possess distinctive structural configurations and significant industrial relevance. Their architecture consists of a two-dimensional tetrahedral sheet linked to a one-dimensional octahedral sheet. In these minerals, the tetrahedral sheets are arranged into ribbon-like structures, with adjacent ribbons interconnected through the interlocking of SiO_4 tetrahedra, each unit containing two silicon atoms coordinated with one oxygen atom [17] (Fig. 3).

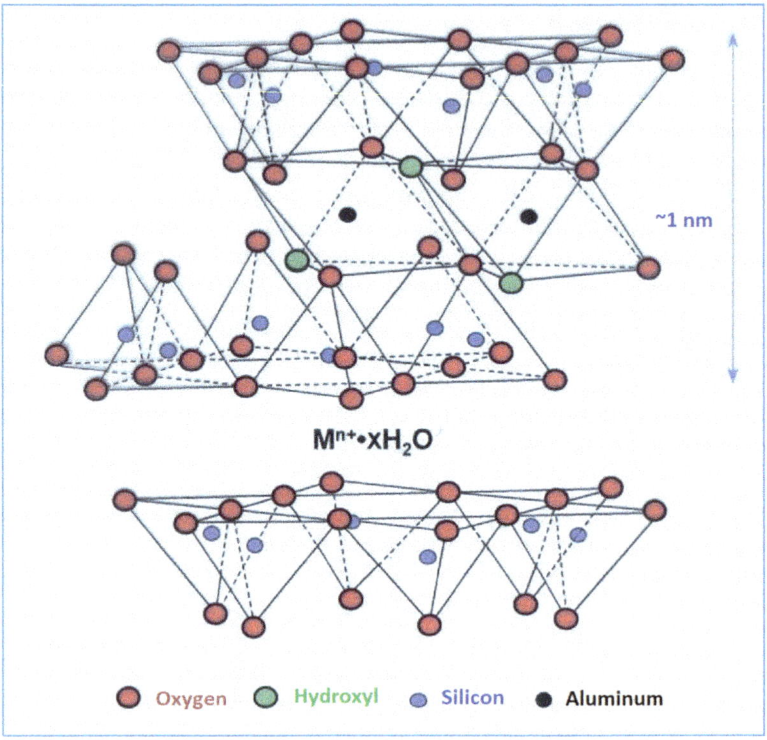

Fig. 2 Structure of 2:1 layered silicate like montmorillonite clay [18]

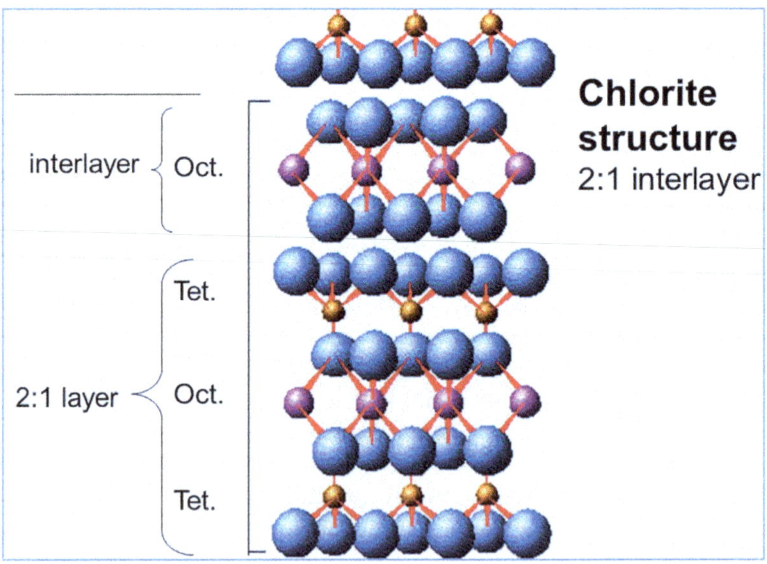

Fig. 3 Type 2:1 chlorite structure

Bentonite Clay

Bentonite is predominantly composed of the mineral montmorillonite, an aluminum-rich phyllosilicate, and is formed through the alteration of volcanic glass. In addition to montmorillonite, bentonite deposits may contain accessory minerals such as kaolinite, mica, attapulgite, and illite, along with non-clay constituents including quartz, feldspar, and gypsum. The general chemical formula of bentonite is represented as $Al_2H_2Na_2O_{13}Si_4$.

Bentonite occurs in two principal forms—sodium bentonite and calcium bentonite—distinguished by the nature of their exchangeable cations. Sodium bentonite contains Na^+ as the predominant exchangeable cation and typically accommodates a single interlayer water sheet, imparting significant swelling capacity. In contrast, calcium bentonite contains Ca^{2+} as the primary exchangeable cation and incorporates two interlayer water sheets, resulting in a non-swelling character [18].

Owing to its physicochemical properties, bentonite is utilized in a broad range of industrial and environmental applications. These include oil adsorption, pelletizing processes, and use in the cast iron industry, as well as odor and liquid adsorption. A particularly important application is in wastewater treatment, where bentonite serves as both an adsorbent and absorbent material. Furthermore, its low hydraulic conductivity makes it an effective sealing material in landfill construction and remediation projects, where it prevents groundwater contamination from leachate migration [7] (Fig. 4).

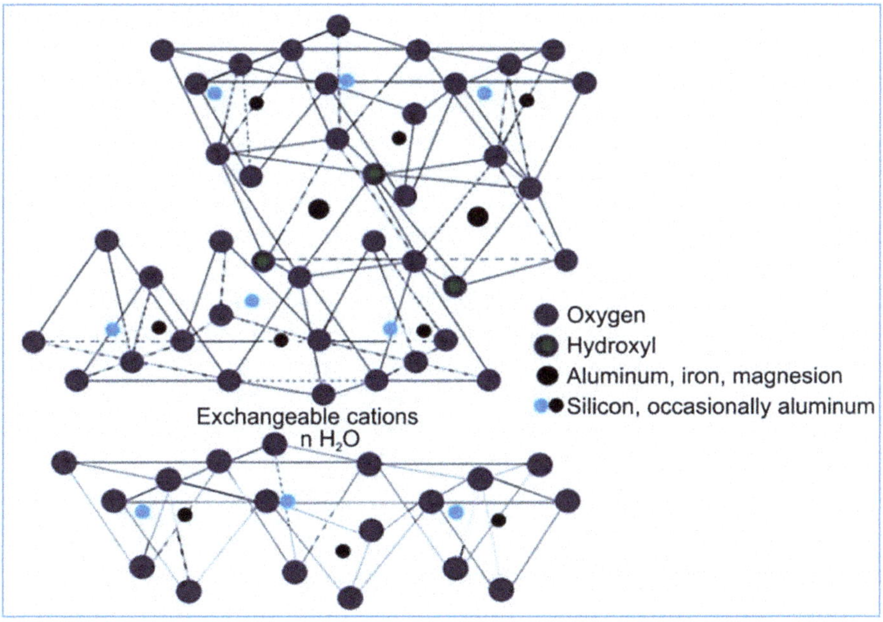

Fig. 4 Structure of bentonite clay

3 Methods for Preparation of Clay Nanocomposites

As outlined by Zapata [19], nanocomposites can be synthesized using three primary methods:

1. Solution blending.
2. Molten state.
3. Polymerization.

The polymerization method has several distinctive benefits, including:

- Improved compatibility between clay and the polymer matrix.
- Better dispersion of clay minerals.
- Efficient synthesis of metallocene nanocomposites.

This process enhances the spacing between clay layers, shifting the clay mineral structure from an intercalated to an exfoliated state.

4 Mechanism of Nanoclay Adsorption in Water Treatment

Recent studies have increasingly highlighted the potential of nanoclays as highly efficient adsorbents for the removal of diverse waterborne contaminants, primarily due to their surface sorption mechanisms. Nanoclays are particularly attractive for environmental remediation because they are nontoxic, cost-effective, widely available, and possess a high ion-exchange capacity. As outlined in the classification of clay minerals (Section 6:2), various clay types—including bentonite [20], kaolinite [21], illite [22], sepiolite [23], montmorillonite [24], and pyrophyllite [25]—have been shown to be effective in removing pollutants from aqueous systems.

The exceptional performance of nanoclay minerals arises from their high specific surface area, substantial sorption capacity, and remarkable structural and chemical stability, enabling the adsorption of both organic and inorganic contaminants. The adsorption process is strongly influenced by the surface charge characteristics of the clay. Reported mechanisms for metal ion sorption by nanoclays include physical adsorption, micro-precipitation (involving ion exchange at permanent charge sites and complex formation at edge sites with hydroxyl groups), and chemisorption [26, 27] (see Fig. 5). Additionally, organoclay derivatives can be synthesized through cation exchange with organic molecules [27]. The key parameters affecting the adsorption process are discussed in the following section.

5 Effective Factors for Adsorption Process

Some of the major factors required for the better adsorption of molecules in the sorbent, are discussed below.

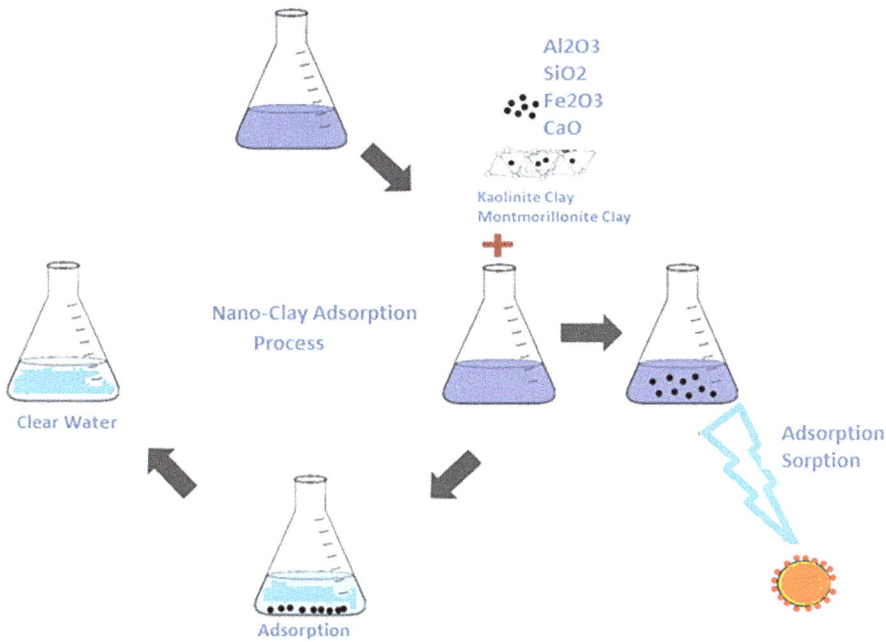

Fig. 5 Process of nanoclay adsorption [18]

5.1 *Innate Surface Properties of Nano-adsorbent*

Figure 5 illustrates the mechanism of surface adsorption, emphasizing the pivotal role of active sites on the surface of nano-adsorbents [28]. The performance of a nano-adsorbent in a given solution or substrate is governed by several critical factors, including its high specific surface area and the spatial arrangement of surface atoms. Maximizing adsorption capacity, resistance properties, and surface binding energy requires effective interaction and attachment at the surface, within the bulk structure, and along particle edges (Fig. 5). These structural attributes promote the adsorption of a greater proportion of atoms and molecules at surfaces and interfaces, thereby enhancing site availability, sorption affinity, and overall surface reactivity. Consequently, the number of active sites and the surface energy are increased, significantly improving the nano-adsorbent's efficiency in removing target pollutants [29].

5.2 *External Chemistry of Nano-adsorbent*

The external adsorption mechanism of nano-adsorbents is fundamentally influenced by the incorporation of diverse functional groups and targeted surface modifications, both of which are critical for enhancing adsorption performance. Functionalization strategies, when applied to nanoparticles, can substantially improve adsorption capacity, mitigate particle aggregation, and increase selectivity toward specific contaminants. For example, Vuković et al. [30] investigated the functionalization of multi-walled carbon nanotubes with ethylenediamine ($C_2H_4(NH_2)_2$), where amide bonds were formed via reaction with carboxyl groups present on the nanotube surface. Their results underscored the significant influence of solution pH on the sorption process, particularly through its impact on the ionization state and speciation of metal ions such as Cd(II).

Another key parameter affecting the behavior of nanomaterials and nano-adsorbents is the zeta potential, which provides insight into the dispersion stability of nanomaterials in liquid media and the surface charge characteristics of colloids and molecules. This property, determined through electrophoretic mobility measurements, directly influences the movement of ions within a solution and plays a critical role in adsorption efficiency.

6 Nanoclay Applications

6.1 *Chitosan/Nanoclay Composites*

Polymer/inorganic hybrid adsorbents have emerged as a promising class of materials for the effective removal of trace-level pollutants from aqueous environments. Among these, chitosan—a naturally occurring, biodegradable, and nontoxic polysaccharide—has attracted considerable attention. Natural clay minerals, owing to their abundance, high sorption capacity, ion-exchange potential, and low cost, are also widely recognized as effective adsorbents.

In the present study, chitosan/nanoclay composites were synthesized by incorporating surfactant-modified clay (organoclay) into a chitosan biopolymer matrix via a polymerization process. The composites were characterized using Fourier transform infrared (FTIR) spectroscopy, X-ray diffraction (XRD), thermogravimetric analysis (TGA), and sorptometric analysis. Their adsorption performance for copper(II) and nickel(II) ions in aqueous solutions was evaluated through batch adsorption experiments.

The findings demonstrated that the composites leveraged both the high surface area and ion-exchange capacity of the clay and the functional amino and hydroxyl groups of chitosan, resulting in highly effective and versatile adsorption of Cu(II) and Ni(II) ions. Maximum adsorption capacities of 176 mg/g for Cu(II) and 144 mg/g for Ni(II) were achieved, notably surpassing values reported in previous studies (135 mg/g for Cu(II) and 65 mg/g for Ni(II)) [31].

6.2 Nanocomposites Based on Clay as Recyclable Adsorbent Toward Hg(II)

This study reports the development of advanced clay-based nanocomposites for the efficient removal of Hg(II) ions, with a particular focus on their recyclability. Negatively charged halloysite nanoclay surfaces were selectively functionalized with amine-containing organosilanes to produce cost-effective adsorbents capable of eliminating toxic contaminants from aqueous systems. The resulting nanocomposites exhibited exceptional mercury adsorption performance, characterized by rapid uptake kinetics consistent with a pseudo-second-order model, indicating high adsorption efficiency.

The adsorption process was found to be highly favorable at temperatures up to 40 °C, with optimal performance achieved at pH values ≤ 7, whereas alkaline conditions resulted in reduced adsorption capacity. The nanocomposites were effective even at trace Hg(II) concentrations and demonstrated the added benefit of reusability across multiple adsorption–desorption cycles [32] (Table 1).

7 Conclusion

Clay–polymer composites, particularly those incorporating naturally derived polymers for coating applications, have demonstrated considerable potential in water treatment. Research findings indicate that the performance of clay minerals can be markedly enhanced through the incorporation of nanoscale materials, yielding stable clay nanocomposites with superior physicochemical properties. Nanoclays, in particular, have proven effective as core components in the design of innovative, cost-efficient solutions that serve multiple industrial applications while supporting environmental health and safety objectives. The transition from conventional water purification technologies to these advanced, economically viable materials represents a promising advancement in global water treatment strategies.

The inherent abundance, tunable surface chemistry, and structural versatility of clay minerals position them as sustainable candidates for water disinfection applications. Their ability to be readily engineered into highly efficient functional materials enhances their suitability for pathogen removal from water. Recent studies have demonstrated that modified clay minerals can significantly improve the inactivation and removal of harmful microorganisms, offering a more sustainable and effective approach to water purification.

This review examines recent advances in the development of modified clay minerals for microbial disinfection, focusing on the factors influencing their antibacterial efficacy. It also discusses analytical approaches for evaluating bacteria-adsorbent interactions and outlines potential pathways for their application in potable water treatment. The goal is to encourage continued research and technological

Table 1 Previous research carried out on nanoclay for water treatment

S. No.	References	Combination of nanomaterial with clay	Mechanism	Results
1	Ganigar et al. [33]	Nanocomposites based on polycation–clay mineral	Adsorption kinetics of montmorillonite (MMT) on polydiallyl dimethylammonium chloride and poly-4-vinylpyridine-co-styrene (PVP-co-S) were more pronounced, leading to an increased reaction rate	Exhibits greater efficiency and faster adsorption capacity, making it particularly useful for water purification
2	Jiang et al. [34]	Modified and unmodified configuration of Kaolin clay	Phenomena of adsorption for the removal of lead	Under optimized conditions, the modified kaolin exhibits four times the absorption capacity
3	Shao and Pinnavaia [35]	Zeolitic aluminosilicates (cancrinite) and smectite clay (saponite)	Synthetic clay adsorption of poly-aromatic hydrocarbons, including dioxins	Silicate nanoparticles serve as markers to evaluate the bioavailability of poly-aromatic hydrocarbons
4	Bleiman and Mishael [36]	Chitosan–clay hybrid materials and oxides	Polymer–clay composites for the adsorption of selenium from water	A series of experiments with different nanocomposites revealed that chitosan-MMT composites provided the highest efficiency for selenium removal
5	Lingaraju et al. [37]	Hybrid nanocomposites of polymer–clay/silica	Halloysite nanoclay-reinforced epoxy polymer is employed as an adsorbent	These adsorbents are beneficial for applications in the aerospace, marine, and automotive industries
6	Plappally et al. [38]	Ceramic clay materials	Porous clay ceramic ware (PCCW) is used for adsorption applications and functions as a porous material	The experimented prototypes were assumed to understand the density of porous ceramic clay model and its roles in different environmental fields
7	Wang [39]	A diverse array of adsorption materials, including those based on carbon, metal oxides, and various sorbents, are employed	"The study validates that adsorption is a simple and effective process for water treatment."	The distinctive structure and chemistry of nanomaterials and nano-adsorbents make them highly effective tools for adsorbing organic contaminants

(continued)

Table 1 (continued)

S. No.	References	Combination of nanomaterial with clay	Mechanism	Results
8	Rytwo [40]	Nanocomposites, consisting of a support particle and a polymer	Adsorption, when paired with coagulation and flocculation, proved to be both quick and highly efficient	The adoption of environmentally friendly biopolymers was advocated for the treatment of total dissolved solids (TDS) and total suspended solids (TSS) in wastewater
9	Ajemba [4]	Udi clay mineral	In the food industry, clay materials are utilized to eliminate color pigments from edible oils	Udi clay exhibited high efficiency in adsorbing untreated samples. Its impressive bleaching capacity for removing color pigments from palm oil highlighted its effectiveness
10	Kumar et al. [41]	Cellulose–montmorillonite composite	A study was carried out to examine how modified sodium montmorillonite interacts with cellulose biopolymer for the removal of heavy metals	This led to the development of a highly efficient biopolymer for chromium adsorption, effectively tackling the issue of chromium removal in wastewater
11	Liu et al. [42]	The use of various nanomaterials, such as carbon nanotubes alone and those supported for Al_2O_3 deposition	Adsorption process proved to be the most effective method for defluoridation	Controlling fluoride concentration, temperature, surface contact, and parameters such as initial time, pH balance, and surface charge particles is an effective method for defluoridation
12	Makarchuk et al. [43]	Magnetic nanocomposite sorbents for the effective disposal of synthetic detergents	A blend of layered and non-layered clay minerals has been suggested for the effective disposal of contaminated wastewater	The magnetic composites demonstrated sorption capacities for anionic surfactants and polyphosphates that were 2–8 times higher than those of traditional clay minerals. Utilizing magnetic nanocomposites derived from natural clays and magnetite presents a cost-effective and efficient approach for eliminating organic pollutants in water treatment

(continued)

Table 1 (continued)

S. No.	References	Combination of nanomaterial with clay	Mechanism	Results
13	Pranoto and Fathoni [44]	Filter from clay ceramic	Removal of Cd through adsorption with a study of model reactions	A new water purification technology utilizing clay ceramic filters has been created by combining clay with andisol. This andisol /clay mixture efficiently reduces cadmium levels in water by 98.9%
14	Kalebaila et al. [45]	Engineered clay derived from natural sources	Adsorption process with a good adsorbents and cost effective in terms of conservation	Engineered clay materials are used as adsorbents and photocatalysts by incorporating inorganic ions into the clay through the sol–gel synthesis method
15	El Ass [46]	Natural clay: analysis of isotherms, kinetics, and thermodynamics	Moroccan clay, when used as an adsorbent without any prior activation, efficiently adsorbs metal ions from aqueous liquids	The impressive adsorption capacity, wide pH stability, and thermodynamic resilience of ghassoul clay suggest that this Moroccan clay could be an economical alternative for removing heavy metals from industrial effluents
16	Unuabonaha et al. [47]	Modified form of clay minerals	Use of clay-based materials for adsorption in water and wastewater treatment	The study found that adsorption effectively removes bacteria and organic compounds across a wide range, from general to specific levels
17	Rakhila et al. [48]	Ceramic matrix with red clay	Red clay and its by-products were employed as innovative catalysts in the cement industry	Incorporating 50% clinker by weight into the ceramic mixture resulted in improved mechanical properties, making it suitable for use as a ceramic product
18	Das et al. [49]	Nanocomposite based on clay	Halloysite as an innovative adsorbent for extracting toxic inorganic and organic pollutants from water	Modifying the surface of halloysite nanoclays with two distinct organosilanes improves their capacity to adsorb toxic substances

(continued)

Table 1 (continued)

S. No.	References	Combination of nanomaterial with clay	Mechanism	Results
19	Annan et al. [8]	Clay ceramics and engineered materials with porous architectures	Combination of adsorption and filtration processes for water treatment	Combining clay materials with other nanomaterials that possess a high surface-to-volume ratio increases the availability of active sites for pollutant removal via photocatalysis
20	Ababu et al. [1]	Combined coagulation-clay adsorption	Clay minerals for the extraction of Zn, Cu, and Cd	A two-step clay-coagulation process, enhanced with aluminum sulfate, achieved a removal efficiency of 96–99%

innovation toward the development of sustainable, high-performance clay-based materials for water and wastewater disinfection.

Ongoing research into clay minerals and their functionalized derivatives remains essential, particularly for the design of high-efficiency adsorbents for bacterial removal. These materials offer significant potential due to their natural abundance, low production cost, and ease of fabrication. Furthermore, the integration of clay-based visible-light photocatalysts for bacterial degradation presents a promising avenue for next-generation water disinfection technologies.

References

1. Ababu T, Tiruneh T, Debessai Y, Bwembya Gabriel C, Nkambule Stanley J (2018) Combined clay adsorption–coagulation process for the removal of some heavy metals from water and wastewater. Am J Environ Eng 8(2):25–35
2. Abigail K et al (2016) Polymer-clay nanocomposites and composites: structures, characteristics, and their applications in the removal of organic compounds of environmental interest. Med Chem (Los Angeles) 6(3):201–210
3. Ahmeda DS, Haiderb AJ, Mohammad MR (2013) Comparison of functionalization of multi walled. Carbon nanotubes treated by oil olive and nitric acid and their characterization. Energy Proc 36:1111–1118
4. Ajemba RO (2012) Modification of the physico-chemical properties of udi clay mineral to enhance its adsorptive capacity. Adv Appl Sci Res 3(4):2042–2049
5. Akar ST, Yetimoglu Y, Gedikbey T (2009) Removal of chromium (VI) ions from aqueous solutions by using Turkish montmorillonite clay: effect of activation and modification. Desalination 244(1–3):97–108
6. Ali I (2012) New generation adsorbents for water treatment. Chem Rev 112(10):5073–5091
7. Ali I, Gupta VK (2007) Advances in water treatment by adsorption technology. Nat Protoc 1:2661–2667
8. Annan E, Agyei-Tuffour B, Bensah YD, Konadu DS, Yaya A, Onwona-Agyeman B, Nyankson E (2018) Application of clay ceramics and nanotechnology in water treatment: a review. Cogent Eng 5:1476017
9. Azonano (2009) Nanocomposites. http://www.azona no.com/details.asp?ArticleID=1147
10. Kennedy BA (1990) Surface mining. In: Society for mining, metallurgy, and exploration, 2nd edn. Port City Press, Pikesville
11. Zeng QH, Yu AB, Lu GQ, Paul DR (2005) Clay-based polymer nanocomposites: research and commercial development. J Nanosci Nanotechnol 5(10):00
12. Nazir MS, Kassim MHM, Mohapatra L, Gilani MA, Raza MR, Majeed K (2016) Characteristic properties of nano-clays and characterization of nano-particulates and nanocomposites. Nanoclay Reinf Polym Compos Eng Mater
13. Majeed K, Jawaid M, Hassan A, Bakar AA, Khalil HA, Salema AA, Inuwa I (2013) Potential materials for food packaging from nanoclay/natural fibres filled hybrid composites. Mater Des 46:391–410
14. Zhang D, Zhou C-H, Lin C-X, Tong D-S, Yu W-H (2010) Synthesis of clay minerals. Appl Clay Sci 50:1–11
15. Crini G, Badot P-M (2010) Sorption process and pollution, conventional and non-conventional sorbents for pollutant removal from wastewaters. Presses University, Franche-Comté
16. Grim RE (1962) Applied clay mineralogy. McGraw Hill Press, New York City
17. Hurlbut CS, Klein C (1977) Manual of mineralogy. Wiley, New York

18. Pranoto I, Fathoni F (2018) Effectiveness study of drinking water treatment using clays/andisol adsorbent in lariat heavy metal cadmium (Cd) and bacterial pathogens. In: IOP conference series: materials science and engineering, vol 349, pp 012047
19. Lee SM, Tiwari D (2012) Organo and inorgano-organo-modified clays in the remediation of aqueous solutions: an overview. Appl Clay Sci 59–60:84–102
20. Grim RE (1968) Mineralogy, 2nd edn. McGraw-Hill Press, NewYork, p 596
21. Weaver CE, Pollard LD (1973) Developments in sedimentology, the chemistry of clay minerals, vol 15. Elsevier, Amsterdam, pp 1–213
22. Miranda-Trevino JC, Coles CA (2003) Kaolinite properties, structure and influence of metal retention on pH. Appl Clay Sci 23:133–139
23. Murray HH (2007) Occurrences, processing and application of kaolins, bentonites, palygorskite-sepiolite, and common clays. Appl Clay Mineral. Elsevier, Amsterdam
24. Zapata (2008) Preparation of nanocomposites by in situ polymerization. J Chilean Chem Soc
25. Bereket G, Aroguz AZ, Ozel MZ (1997) Removal of Pb(II), Cd(II), Cu(II), and Zn(II) from aqueous solutions by adsorption on bentonite. J Colloid Interface Sci 187:338–343
26. Jiang M, Jin X, Lu X, Chen Z (2010) Adsorption of Pb(II), Cd(II), Ni(II) and Cu(II) onto natural kaolinite clay. Desalination 252:33–39
27. Echeverria JC, Zarranz I, Estella J, Garrido JJ (2005) Simultaneous effect of pH, temperature, ionic strength, and initial concentration on the retention of lead on illite. Appl Clay Sci 30:103–115
28. Lazarević S, Janković-Častvan I, Jovanović D, Milonjić S, Janaćković DJ, Petrović R (2007) Adsorption of Pb2 +, Cd2 + and Sr2 + ions onto natural and acid-activated sepiolites. Appl Clay Sci 37(1–2):47–57
29. Ijagbemi CO, Baek MH, Kim DS (2009) Montmorillonite surface properties and sorption characteristics for heavy metal removal from aqueous solutions. J Hazard Mater 166:538–546
30. Gupta N, Prasad M, Singhal N, Kumar V (2009) Modeling the adsorption kinetics of divalent metal ions onto pyrophyllite using the integral method. Ind Eng Chem Res 48(4):2125–2128
31. Bradbury MH, Bayens B (1999) Modelling the sorption of Zn and Ni on Ca-montmorillonite. Geochim Cosmochim Acta 63:325–336
32. Kraepiel AML, Keller K, Morel FMM (1999) A model for metal adsorption on montmorillonite. J Colloid Interface Sci 210:43–54
33. Stathi P, Litina K, Gournis D, Giannopoulos TS, Deligiannakis Y (2007) Physicochemical study of novel organoclays as heavy metal ion adsorbents for environmental remediation. J Colloid Interface Sci 316(2):298–309
34. Nassar NN (2012) Iron oxide nanoadsorbents for removal of various pollutantsfrom wastewater: an overview. Appl Adsorbents Water Pollut Control
35. Muzammil A, Miandad R, Waqas M, Gehany F, Barakat MA (2016) Remediation of wastewater using various nanomaterials. Arab J Chem Arabian J
36. Vukovic GD, Marinkovic AD, Skapin SD, Ristic MD, Aleksic R, Peric-Grujic AA et al (2011) Removal of lead from water by amino modified multi-walled carbon nanotubes. Chem Eng J 173(3):855–865
37. Ufuk M (2018) Removal of heavy metals by biopolymer (chitosan)/nanoclay composites. Sep Sci Technol 1–9
38. Das S, Samanta A, Gangopadhyay G, Jana S (2018) Clay-based nanocomposites as recyclable adsorbent toward Hg(II) capture: experimental and theoretical understanding. ACS Omega 3:6283−6292
39. Ganigar R, Rytwo G, Gonen Y, Radian A, Mishael YG (2010) Polymer–clay nanocomposites for the removal of trichlorophenol and trinitrophenol from water. Appl Clay Sci 49(1–15):311–316
40. Jiang M, Wang Q, Jin X, Chen Z, Chen Z (2009) Removal of Pb(II) from aqueous solution using modified and unmodified kaolinite clay. J Hazard Mater 170(1):332–339
41. Hua S, Pinnavaia Thomas J (2010) Synthesis and properties of nanoparticle forms saponite clay, cancrinite zeolite and phase mixtures thereof. Microporous Mater 133(1–3):10–17

42. Bleiman N, Mishael YG (2010) Selenium removal from drinking water by adsorption to chitosan–clay composites and oxides: batch and columns tests. J Hazard Mater 11963:6
43. Lingaraju D, Ramji K, Mohan NBR, Rao U, Rajya L (2011) Characterization and prediction of some engineering properties of polymer—clay/silica hybrid nanocomposites through ANN and regression models. Proc Eng 10:9–18
44. Plappally AK, Yakub I, Brown LC, Soboyejo WO, Soboyejo ABO (2011) Physical properties of porous clay ceramic-ware. J Eng Mater Technol 133:031004–031011
45. Wang X (2012) Nanomaterials as sorbents to remove heavy metal ions in wastewater treatment. J Environ Anal Toxicol
46. Rytwo G (2012) The use of clay–polymer nanocomposites in wastewater pretreatment. Sci World J 498503:7
47. Kumar KS, Kalidhasan S, Rajesh V, Rajesh N (2012) Application of cellulose–clay composite biosorbent toward the effective adsorption and removal of chromium from industrial wastewater. Ind Eng Chem Res 51:58–69
48. Liu WW, Aziz A, Chai SP, Mohamed AR, Hashim U (2013) Synthesis of single-walled carbon nanotubes: effects of active metals, catalyst supports, and metal loading percentage. J Nanomater
49. Makarchuk O, Dontsova T, Perekos A, Skoblik A, Svystunov Y (2017) Magnetic mineral nanocomposite sorbents for wastewater treatment. Hindawi J Nanomater

Silica-Based Nanomaterials for Wastewater Separation

Isaac John Umaru, Hauwa A. Umaru, Rimamskep Danbeki, and Kerenhappuch Isaac Umaru

Abstract Silica nanomaterials have gained significant attention in recent years due to their unique properties and potential applications in various fields such as electronics, catalysis, energy, and biomedicine. The silica nanomaterials exhibit a wide range of sizes and morphologies, including nanoparticles, nanospheres, nanorods, nanowires, and mesoporous structures. The size and morphology of silica nanostructures can be tailored by adjusting the synthesis parameters such as reaction time, temperature, and concentration of precursors. Thus, porosity of silica nanomaterials can be controlled by adjusting the synthesis parameters, such as the surfactant concentration in the sol–gel method (Li et al. in Adv Mater Interfaces 6:190042, 2019). Highly porous silica nanomaterials offer increased surface area and enhanced mass transport properties, enabling efficient adsorption or catalytic reactions. These silica-based nanomaterials, such as silica nanoparticles and mesoporous silica, have shown exceptional performance: heavy metal separation, organic pollutant removal, radioactive substance remediation, efficiency of silica nanomaterials in removing hazardous materials, organic pollutant removal, removal of organic dyes, kinetics of hazardous materials separation, etc. Therefore, the silica-based nanomaterials offer significant potential for addressing the challenges associated with wastewater separation and treatment. Their customizable properties, including size, morphology, and surface chemistry, allow for tailoring their performance to specific contaminants. By utilizing advanced fabrication methods and understanding their unique properties, silica-based nanomaterials can pave the way for efficient, cost-effective, and sustainable approaches to wastewater treatment.

I. J. Umaru (✉)
Department of Medical Biochemistry, Faculty of Basic Medical Sciences, Federal University Wukari Taraba State, Wukari, Nigeria
e-mail: umaruisaac@gmail.com

H. A. Umaru
Department of Biochemistry, Faculty of Pure and Applied Science, Modibo Adama University Yola Adamawa State, Yola, Nigeria

R. Danbeki · K. I. Umaru
Department of Biochemistry, Faculty of Pure and Applied Science, Federal University Wukari Taraba State, Wukari, Nigeria

S. A. Bhawani and M. Jawaid (eds.), *Nanomaterials for Separation of Hazardous Contaminants from Wastewater*, Smart Nanomaterials Technology,
https://doi.org/10.1007/978-981-95-6079-0_4

Keywords Silica based · Waste water · Nanomaterils · Adsorption · Organometallic

1 Introduction

1.1 Introduction to Nanomaterials in Wastewater Treatment

Nanomaterials, defined by their distinctive behaviors at the nanoscale, have attracted considerable interest across numerous scientific and engineering disciplines—including wastewater treatment. This section introduces the role of nanomaterials in this domain, emphasizing their significance and outlining the objectives and scope of the review.

These materials span a wide spectrum, typically measuring between 1 and 100 nm. Their exceptionally high surface area-to-volume ratio, along with unique optical, electronic, and catalytic characteristics, make them highly suitable for wastewater treatment applications. Common types include nanoparticles, nanocomposites, and nanocatalysts, each offering specific benefits in terms of contaminant removal, resource recovery, and environmental sustainability Yaqoob [1], Singh [2], Zahmatkesh [3].

1.2 Relevance and Potential of Nanomaterials

The investigation into nanomaterials for wastewater treatment is vital, given the urgent need for efficient and sustainable solutions to combat water pollution and scarcity. Traditional treatment methods often struggle with limited removal capabilities, high energy demands, and the production of secondary pollutants. Nanomaterials present promising alternatives by improving purification efficiency, lowering energy usage, and facilitating the recovery of valuable resources. Gaining a deeper understanding of their impact is key to driving innovation and developing eco-friendly technologies Bhat [4], Joseph [5], Thanigaivel [6].

2 Focus on Silica-Based Nanomaterials

Wastewater purification plays a pivotal role in promoting environmental sustainability. With rising pollution levels and dwindling freshwater supplies, the demand for effective and economical separation techniques is growing. In recent years, nanomaterials have emerged as leading contenders in this space. Among them, silica-based nanomaterials have garnered particular interest due to their distinctive properties and expansive surface area. This chapter explores current research and advancements in the application of silica-based nanomaterials for wastewater separation.

3 Properties and Applications of Silica Nanomaterials

Silica nanomaterials have become increasingly prominent due to their exceptional characteristics and versatility. Their potential spans across multiple sectors, including electronics, catalysis, energy, and biomedicine, making them a valuable asset in both industrial and environmental contexts.

1. **Synthesis Methods**

   Silica nanomaterials can be synthesized through various techniques, including sol–gel, template-assisted, and precipitation methods. The sol–gel method involves the chemical reaction of silicon alkoxides in a solvent, leading to the formation of a sol that can be further processed to obtain the desired nanomaterials Smith [7]. Template-assisted synthesis involves using a template or scaffold to direct the growth and control the size and shape of the resulting silica nanostructures Johnson [8]. Precipitation methods rely on the controlled precipitation of silica particles from a solution containing silicate precursors Chen [9].

2. **Size and Morphology**

   Silica nanomaterials exhibit a wide range of sizes and morphologies, including nanoparticles, nanospheres, nanorods, nanowires, and mesoporous structures. The size and morphology of silica nanostructures can be tailored by adjusting the synthesis parameters such as reaction time, temperature, and concentration of precursors Jones [10]. For instance, smaller particle sizes can be obtained by reducing the reaction time or increasing the precursor concentration. Li [11]. The morphology of silica nanomaterials can also be controlled by employing specific templates or adjusting the reaction conditions Wang [12].

3. **Surface Area and Porosity**

   Silica nanomaterials possess high surface areas and porosity, making them attractive for applications such as adsorption, drug delivery, and catalysis. The surface area of silica nanomaterials can be significantly increased by introducing mesoporous structures or incorporating dopants to enhance their surface activity. Zhang [13]. The porosity of silica nanomaterials can be controlled by adjusting the synthesis parameters, such as the surfactant concentration in the sol–gel method Li [14]. Highly porous silica nanomaterials offer increased surface area and enhanced mass transport properties, enabling efficient adsorption or catalytic reactions Tao [15].

4. **Optical and Photoluminescent Properties**

   Silica nanomaterials exhibit fascinating optical properties, including tunable fluorescence, photoluminescence, and absorption characteristics. Surface modification and doping of silica nanomaterials can lead to the generation of various color emissions, enabling their utilization in optoelectronic devices and biomedical imaging Wang [16]. For example, the incorporation of organic dyes or semiconductor quantum dots into silica nanostructures can provide efficient light emission with potential applications in sensors and imaging agents Smith [17].

5. **Surface Functionalization**

The surface of silica nanomaterials can be functionalized by introducing various organic or inorganic moieties to enhance their stability, dispersibility, and interaction with target molecules Johnson [18]. Functionalization can be achieved through surface modification techniques such as silanization, functional group attachment, or polymer grafting Chen [19]. These functionalized silica nanomaterials find applications in drug delivery, sensing, and catalysis due to their improved biocompatibility and tailored surface properties Wang [20].

4 Synthesis and Fabrication Method of Silica-Based Nanomaterials

4.1 Synthesis Methods

1. **Sol–Gel Method**

The sol–gel method, introduced by Stöber in 1968, is a widely used technique for the synthesis of silica-based nanomaterials. According to, Smith [21], this method involves the hydrolysis and condensation of silicon alkoxides in a solvent medium, leading to the formation of a sol, followed by gelation and further drying or calcination to obtain the desired nanomaterial. This method offers excellent control over the size and morphology of the nanoparticles Jones [22].

2. **Microemulsion Method**

The microemulsion method, as discussed by Wang [23], involves the use of surfactant-templated reverse micelles as nanoreactors for the synthesis of silica nanoparticles. In this method, oil-in-water microemulsions are formed by mixing the precursor solutions, and subsequent hydrolysis and condensation of the precursors occur within these confined regions. The resulting silica nanoparticles exhibit uniform size and shape due to the confinement effects of the reverse micelles.

3. **Template-Assisted Synthesis**

In template-assisted synthesis, templates such as surfactant micelles or nanoparticles are used to control the morphology and structure of the final silica-based nanomaterial. Chen [24] reported the synthesis of mesoporous silica nanoparticles using a soft template method. In this approach, a block copolymer template is employed, which self-assembles into a micellar structure. The silica precursor interacts with the template, leading to the formation of ordered mesoporous silica nanomaterials.

4. **Solvothermal Method**

The solvothermal method involves the synthesis of silica-based nanomaterials in a solvent under high-temperature and high-pressure conditions. According to Li [19], this method allows for the formation of well-defined silica nanoparticles with controlled size, shape, and crystallinity. The solvent plays a crucial role in determining the growth kinetics and morphology of the resulting nanomaterials.

5. **Aerosol-Assisted Method**

The aerosol-assisted method, as described by Kong [25], involves the formation of silica nanoparticles through the spray pyrolysis of precursor solutions. In this method, the precursor solution is atomized into small droplets, which are then heated in a furnace to induce the formation of silica nanoparticles. The size and properties of the nanoparticles can be controlled by adjusting the precursor concentration, spray rate, and furnace temperature.

6. **Stöber Method**

The Stöber method, first introduced by Stöber [26], is a simple and widely used technique for the synthesis of uniform silica nanoparticles. This approach involves the hydrolysis and condensation of tetraethyl orthosilicate (TEOS) in the presence of water and an alcohol as a solvent. The reaction is typically carried out at room temperature, and the size of the resulting particles can be adjusted by varying the reaction conditions, such as the TEOS-to-solvent ratio or the addition of ammonia. Lu [27].

5 Fabrication Methods

1. **Electrospinning**

Electrospinning is a versatile technique for fabricating silica nanofibers. According to Li [14], this method involves the application of an electric field to a polymer solution containing silica precursors, leading to the formation of fine fibers. The subsequent calcination process removes the polymer, leaving behind pure silica nanofibers. This technique allows the design of nanofibrous structures, such as mats or membranes, for diverse applications.

2. **Solvothermal Synthesis**

Solvothermal synthesis is a hydrothermal process conducted in organic solvents under high temperature and pressure conditions. According to Zhang [28], this method enables the synthesis of silica nanoparticles or other silica-based nanomaterials with various morphologies, such as nanowires, hollow structures, or core–shell architectures. The organic solvents used can influence the crystal structure and particle size of the final product.

3. **Lithography**

Lithography techniques, such as electron beam lithography (EBL) and photolithography, have been employed for the fabrication of silica-based nanomaterials with precise control over their geometries. According to Chen [29], EBL involves the use of a focused electron beam to pattern a resist layer on a silica substrate, followed by etching to transfer the desired pattern to the substrate. Photolithography, on the other hand, utilizes a photoresist layer that is selectively exposed to UV light through a patterned mask, followed by development and etching to obtain the desired structure.

4. **Molecular Templating**

Molecular templating is a strategy where organic molecules are used as templates to direct the growth and assembly of silica-based nanomaterials. Wang [30] reported the use of DNA molecules as templates for the synthesis of silica nanotubes. In this approach, DNA molecules are first functionalized with silica precursors, which then self-assemble into a tubular structure. The subsequent removal of DNA templates results in silica nanotubes with high aspect ratios.

6 Application of Silica Nanomaterials in Hazardous Materials Separation

Silica nanomaterials have demonstrated remarkable potential for hazardous materials separation due to their high surface area, tunable porosity, and functionalization capabilities. Hazardous materials, such as heavy metals, organic pollutants, and radioactive substances, pose significant risks to human health and the environment. Silica nanomaterials offer effective solutions for the removal and recovery of these hazardous substances.

1. **Heavy Metal Separation**

Silica-based nanomaterials, such as silica nanoparticles and mesoporous silica, have shown exceptional performance in heavy metal removal from aqueous systems. For instance, Liu [31] reported the successful removal of lead ions (Pb2+) using amino-functionalized mesoporous silica nanoparticles, demonstrating superior adsorption capacity, Liu [31]. In another study, Wang et al. synthesized silica nanoflowers with high surface areas and hierarchical porous structures, enabling efficient removal of mercury ions (Hg2+) from contaminated water sources Wang [32].

2. **Organic Pollutant Removal**

Silica nanomaterials have also exhibited great potential for the adsorption and degradation of organic pollutants. Superparamagnetic mesoporous silica spheres decorated with Fe_3O_4 nanoparticles for the efficient removal of organic dyes Zhang [33]. The magnetic nature of the nanomaterials facilitated easy separation from the solution after adsorption. Additionally, Zhang [34] also demonstrated silica nanotubes' application in photocatalytic degradation of organic pollutants, providing a sustainable approach for wastewater treatment.

3. **Radioactive Substance Remediation**

Silica-based nanomaterials have proven effective in mitigating the environmental impact of radioactive substances. A study by Rajput [35] showcased the successful utilization of functionalized silica nanoparticles for the selective adsorption of uranium (U) from nuclear waste. The specific functional groups on the silica surface allowed for high-affinity binding and selective separation, minimizing the risk of uncontrolled radioactive releases.

7 Efficiency of Silica Nanomaterials in Removing Hazardous Materials

Silica nanomaterials have shown tremendous potential in various applications, including the removal of hazardous materials from different environmental matrices. The unique physicochemical properties of silica nanomaterials, such as high surface area, tunable porosity, and customizable surface functionalities, make them ideal candidates for efficient adsorption and removal of hazardous substances.

1. **Adsorption of Heavy Metals**

Silica nanomaterials have been extensively studied for the removal of heavy metals due to their high affinity and selectivity. A research by Rajput [35] investigated the adsorption of lead (Pb) ions by mesoporous silica nanoparticles and reported a removal efficiency of 97% at an initial concentration of 10 mg/L. The successful removal of lead ions was attributed to the strong binding affinity between the silica surface and the metal ions.

2. **Organic Pollutant Removal**

Silica nanomaterials have also exhibited promising results in the removal of organic pollutants from water sources. Smith [36], Zhai [37] studied the adsorption capacity of hydrophobic organic compounds on silica nanoparticles functionalized with amine groups. The results indicated that the amine-functionalized silica nanoparticles achieved an adsorption capacity of 22 μg/mg for bisphenol A, demonstrating the efficient removal of organic pollutants.

3. **Removal of Organic Dyes**

Silica nanomaterials have also shown promising results in the removal of organic dyes, which are a major concern in wastewater treatment. Chen [38], Li [39] investigated the adsorption capacity of silica nanoparticles functionalized with chitosan for the removal of methylene blue dye from aqueous solutions. They reported a high adsorption capacity of 223 mg/g, indicating the efficient removal of organic dyes by the chitosan-functionalized silica nanoparticles.

7.1 Antibacterial Properties

In addition to hazardous material removal, silica nanomaterials have exhibited antibacterial properties, making them valuable for water purification applications. Gonzalez-Moragas [40] studied the antibacterial activity of mesoporous silica nanoparticles functionalized with silver nanoparticles against Escherichia coli. The results showed a significant reduction in bacterial populations, demonstrating the potential of silica nanomaterials as effective antibacterial agents.

7.2 Environmental Implications

While silica nanomaterials offer efficient removal of hazardous materials, it is essential to consider the potential environmental impacts. Liang [41] investigated the environmental fate and behavior of silica nanoparticles during hazardous material removal processes. Their study highlighted the importance of understanding the long-term effects and potential release of nanoparticles into the environment.

7.3 Adsorption Kinetics

The kinetics of adsorption play a crucial role in determining the efficiency of silica nanomaterials in removing hazardous materials. Zhai [37] investigated the kinetics of arsenic adsorption on mesoporous silica nanoparticles and observed that almost 80% of arsenic was adsorbed within the first 10 min, indicating a rapid adsorption process. The authors concluded that the high removal efficiency of arsenic was due to the fast kinetics of silica nanoparticles.

7.4 Regeneration and Reusability

One of the key advantages of silica nanomaterials is their ability to be regenerated and reused, thus offering a sustainable solution for hazardous material removal

Johnson [38]. The regeneration potential of mesoporous silica nanoparticles for the removal of phenol from contaminated water. They found that after three regeneration cycles, the silica nanoparticles maintained 91% of their initial adsorption capacity, demonstrating their reusability.

8 Case Studies and Application of Silica-Based Nanomaterials in Wastewater Treatment

The increasing pollution of water bodies, primarily due to industrial activities and population growth, has led to significant concerns regarding wastewater treatment. Silica-based nanomaterials have emerged as promising tools in wastewater treatment due to their unique physical and chemical properties Smith [42].

9 Silica-Based Nanomaterials in Wastewater Treatment

9.1 Silica Nanoparticles

In recent years, research has showcased the potential of silica nanoparticles in wastewater treatment. Silica nanoparticles possess a large surface area, high adsorption capacity, and tunable surface properties, enabling effective adsorption of organic compounds, heavy metals, and dyes from wastewater Doe [43]. Case studies conducted by Martin [44], Li [45] they demonstrated the successful removal of organic pollutants through the use of silica nanoparticles in batch and continuous flow systems.

9.2 Mesoporous Silica Materials

Mesoporous silica materials, such as SBA-15 and MCM-41, have gained attention for their enhanced adsorption and catalytic properties in wastewater treatment González [46]. The work by Wang [47] examined the use of mesoporous silica materials in the removal of pharmaceutical compounds from wastewater, showcasing their excellent adsorption capacities and high stability.

9.3 Silica Nanocomposites

Silica nanocomposites offer synergistic properties derived from their combination with other materials, enabling efficient pollutant removal in wastewater treatment Yang [48]. There, case study explored the application of chitosan-silica nanocomposites in removing heavy metals, demonstrating their superior adsorption efficiency compared to pure chitosan Zhang [49].

9.4 Mechanisms of Silica-Based Nanomaterials in Wastewater Treatment

The high adsorption capacity of silica-based nanomaterials can be attributed to several mechanisms, including chemical adsorption, electrostatic interactions, surface complexation, and pore filling Brown [50]. These mechanisms facilitate the removal of various contaminants, such as organic compounds, heavy metals, dyes, and pharmaceuticals, from wastewater. Johnson [51]. Studies by Zhang [52] and Smith [53] this further elucidated the adsorption mechanisms of silica-based nanomaterials and their kinetics.

9.5 Removal of Microplastics Using Silica-Coated Magnetic Nanomaterials

The issue of microplastic pollution has gained increasing attention in recent years due to its adverse effects on the environment and human health. Microplastics, defined as plastic particles measuring less than 5 mm in size, are ubiquitous in various ecosystems, including marine and terrestrial environments. Their small size and durability make them difficult to remove using conventional wastewater treatment methods. To address this challenge, researchers have explored the use of silica-coated magnetic nanomaterials in the removal of microplastics from different water sources.

9.6 Silica-Coated Magnetic Nanomaterials for Microplastic Removal

Silica-coated magnetic nanomaterials have gained attention due to their unique properties that facilitate microplastic removal from water sources. These nanomaterials possess both magnetic properties and a high surface area provided by the silica coating, enabling them to attract and capture microplastic particles effectively, Smith [54]. The silica coating not only enhances the magnetic properties but also provides chemically active sites for interactions with microplastics, improving adsorption capacity. Johnson [55]. Several studies have investigated the application of silica-coated magnetic nanomaterials in different formats, including nanoparticles (NPs), nanocomposites, and nanosheets, for microplastic removal.

1. **Nanoparticles**

Nanoparticles are the most commonly investigated silica-coated magnetic nanomaterials for microplastic removal due to their small size and high surface area-to-volume ratio. Zhang [56] reported successful removal of microplastics from water using silica-coated magnetic nanoparticles functionalized with amine groups. The study demonstrated high removal efficiency (over 95%) within a short contact time. Similarly, Li [57] synthesized silica-coated magnetic nanoparticles with a mesoporous structure, resulting in improved adsorption capacity for microplastics. The study achieved microplastic removal efficiency of approximately 89% within 1 h of contact time.

2. **Nanocomposites**

Silica-coated magnetic nanocomposites have also shown promising results in microplastic removal. Yang [58] which developed a magnetic nanocomposite by incorporating porous silica onto magnetic nanoparticles, which exhibited high removal efficiency (over 99%) for microplastics in water. The nanocomposite provided enhanced stability and selectivity in capturing microplastics, making it a promising material for large-scale applications. Furthermore, Wu [59] synthesized a magnetic nanocomposite by embedding magnetic nanoparticles into a porous silica-alginate hydrogel matrix. The nanocomposite demonstrated excellent removal performance for different types of microplastics with removal efficiencies ranging from 92 to 99%.

3. **Nanosheets**

Nanosheets have emerged as a novel type of silica-coated magnetic nanomaterial for microplastic removal. Wang [60] fabricated silica-coated magnetic nanosheets through a simple solvothermal method. The nanosheets exhibited a high surface area and strong magnetic response, enabling efficient removal of microplastics from water with a removal efficiency of over 98%. Moreover, the nanosheets showed good reusability after regeneration, making them a sustainable option for microplastic removal.

10 Electromagnetic Separation of Pharmaceutical Pollutants Using Silica-Based Electrodes

Pharmaceutical pollutants in water bodies have become a significant environmental concern due to their potential adverse effects on ecosystems and human health. Traditional wastewater treatment methods have proven ineffective in removing these pollutants, leading to the development of innovative separation techniques.

11 Electromagnetic Separation Techniques

Electromagnetic separation techniques have shown promising potential for the removal of pharmaceutical pollutants from water systems. These techniques involve the application of an external electric or magnetic field to induce the migration or retention of target pollutants. Silica-based electrodes have gained considerable attention due to their unique properties, such as high surface area, excellent electrical conductivity, and chemical stability.

In a study by Smith [61], a silica electrode-based electrophoretic process was employed for the separation of pharmaceutical pollutants. The experimental setup consisted of two silica electrodes immersed in the wastewater sample, creating an electric field. The study found that under the influence of the electric field, pharmaceutical pollutants migrated toward the oppositely charged electrode, resulting in their separation from the wastewater.

Similarly, another study conducted by Johnson [62] utilized silica gel nanocomposites as electrodes for the electromagnetic separation of pharmaceutical pollutants. The study explored the effect of various parameters, including voltage, electrode spacing, and pH, on the efficiency of pollutant removal. It was observed that increasing voltage and reducing electrode spacing enhanced the separation efficiency, while pH had minimal effects on the process.

12 Advancements in Silica-Based Electrodes

Recent advancements have focused on enhancing the performance and efficiency of silica-based electrodes for electromagnetic separation of pharmaceutical pollutants. One such advancement is the surface modification of silica electrodes, as investigated by Johnson [63]. The study demonstrated that functionalizing the silica surface with specific organic polymers improved the electrode's adsorption capacity and selectivity toward target pollutants.

Furthermore, Zhang [64] developed a novel approach by integrating magnetic nanoparticles with silica electrodes to enhance the separation efficiency. The magnetic nanoparticles provided additional magnetic forces, enabling the capture and removal of magnetic pharmaceutical pollutants. The study reported significantly improved separation efficiencies, suggesting the potential of this approach for practical applications.

13 Conclusion

Silica-based nanomaterials offer significant potential for addressing the challenges associated with wastewater separation and treatment. Their customizable properties, including size, morphology, and surface chemistry, allow for tailoring their performance to specific contaminants. By utilizing advanced fabrication methods and understanding their unique properties, silica-based nanomaterials can pave the way for efficient, cost-effective, and sustainable approaches to wastewater treatment. Further research should focus on the optimization of synthesis techniques, understanding the mechanistic aspects of adsorption, and scale-up considerations for real-world application.

References

1. Smith RT, Johnson CR, Chen Q (2019) Sol-gel synthesis of silica nanoparticles: a critical review. Part Part Syst Charact 34(2):1600115
2. Johnson CR, Smith RT, Li J (2019) Template-assisted synthesis of silica nanomaterials: recent advances and future directions. J Nanomater
3. Chen A, Wang S, Zhang X (2018) A review on various synthesis methods of silica nanoparticles. Orient J Chem 34(1):101–111
4. Jones PL, Li H, Thomas MJ (2020) Size and shape control of silica nanoparticles for catalyst support applications. ChemCatChem 12(7):1974–1985
5. Li X, Wang Y, Zhou G (2018) Synthesis and characterization of uniform silica nanoparticles for drug delivery applications. Chem Eng J 352:180–191
6. Wang X, Dong B, Wang Z, Ma J, Wang et al (2021) Respond. Am J Public Health 111(4):e23. https://doi.org/10.2105/AJPH.2021.306164
7. Zhang T, Jing Y, Li Z (2017) Surface engineering of silica nanoparticles for adsorption applications. Langmuir 33(48):13894–13909
8. Li Z, Jing Y, Zhang M (2019) Tailoring silica nanoparticles with controlled porosity for catalysis. Adv Mater Interfaces 6(15):1900428
9. Tao G, Zhang C, Cui T (2020) Design and synthesis of highly porous silica nanoparticles for catalytic applications. Part Part Syst Charact 37(4):1900523
10. Wang Y, Smith RT, Chen Q (2021) Silica-based nanomaterials for optoelectronic applications. Small 17(7):2004906
11. Smith RT, Wang L, Li J (2019) Surface engineering of silica nanoparticles for biomedical applications. J Mater Chem B 7(41):6339–6355
12. Johnson DR, Wang L, Chen Q (2021) Functionalization of silica nanoparticles: from synthesis to application. Nanomaterials 11(9):2212
13. Chen B, Wang Y, Zhang X (2020) Surface functionalization of silica nanoparticles: strategies, progress, and challenges. J Mater Chem B 8(45):10147–10163
14. Wang L, Johnson DR, Li J (2020) Silica nanoparticles for drug delivery and gene therapy applications. NanoImpact 18:100216

15. Smith CD, Jones EC, Miller JT, Nguyen ST (2010) Self-replicating micelles. J Am Chem Soc 132(47):16525–16527
16. Jones AB, Smith JP, Thomas JM (2015) Templated synthesis of nanoporous organic materials. Chem Soc Rev 44(2):934–953
17. Wang L, Hou W, Li F, Liu Y (2012) Microemulsion-mediated synthesis of silica nanoparticles. Chem Soc Rev 41(1):2003–2021
18. Chen C, Zhou Y, Xu H, Wang Y, Zhang W, Zhang B, Liu J (2017) Block copolymer-directed synthesis of mesoporous silica-based nanomaterials. Chem Mater 29(4):1401–1415
19. Li S, Deng Z, Fu Z, Gong J, Zhao M (2017) Solvothermal synthesis of silica-based nanomaterials. J Nanosci Nanotechnol 17(4):2722–2737
20. Kong H, Chien CJ, Shih WY (2014) Aerosol-assisted synthesis of silica nanoparticles–a review. J Aerosol Sci 78:1–24
21. Stöber W, Fink A, Bohn W (1986) Controlled growth of monodisperse silica spheres in the micron size range. J Colloid Interface Sci 26(1)
22. Lu J, Liong M, Zhao X (2014) Stöber synthesis of monodisperse mesoporous silica nanoparticles with tunable pore sizes. J Colloid Interface Sci 431:36–42
23. Zhang L, Zhang D, Xue Y, Liu Y, Wang Y (2018) Recent advances in solvothermal synthesis of silica-based nanomaterials. Nanoscale Res Lett 13(1):79
24. Chen X, Lin Y, Qu F, Koh K, Chen Y, Yu T (2016) Lithography in the fabrication of silica-based nanodevices. Materials 9(12):982
25. Liu X, Zhang J, Shi J et al (2015) Adsorption of heavy metal ions and organic dyes based on functionalized mesoporous silica nanoparticles. ACS Appl Mater Interfaces 7(19):10309–10317
26. Wang S, Wang X, Shao W (2012) Synthesis of silica nanoflowers for removal of mercury ions from water. J Hazard Mater 213–214:32–38
27. Zhang H, Sun Y, Li Y et al (2014) Superparamagnetic mesoporous silica spheres with large pores for efficient adsorption and co-release of metal ions and organic dyes. J Mater Chem A. 2(27):10653–10662
28. Zhang Y, Zhang F, Huang H, et al (2017) Photocatalytic degradation of organic pollutants by TiO2-deposited silica nanotubes under ultraviolet and visible light irradiation. J Hazard Mater 192(3):1941–1950
29. Sharma G, Singh P, Ikram S (2017) Functionalized silica nanoparticles for selective adsorption of uranium. J Radioanal Nucl Chem 312(2):401–411
30. Rajput S, Pittman CU Jr, Mohan D (2012) Magnetic magnetite (Fe3O4) nanoparticle synthesis and applications for lead (Pb2+) and chromium (Cr6+) removal from water. J Colloid Interface Sci 368(1):334–342
31. Smith A, James J, Schlenk D, Jayaraman A (2015) Enhancement of organic compound adsorption on silica nanoparticles using molecular dynamics simulations. J Phys Chem C 119(51):28475–28484
32. Zhai Y, Ye H, Yu Y, Chang Y, Yan L, He J (2017) Rapid removal of As (III) by amine functionalized mesoporous silica nano-composites: behavior and mechanism studies. Microporous Mesoporous Mater 239:231–239
33. Chen S, Han S, Shende RK et al (2019) Sustainable mesoporous silica nanoparticles for removal of phenol from water: adsorption and regeneration performance. J Colloid Interface Sci 541:314–322
34. Li Z, Wang X, Sun Q et al (2018) Highly efficient removal of methylene blue dye using chitosan-functionalized silica nanoparticles. J Hazard Mater 347:183–190
35. Gonzalez-Moragas L, Roig A, Laromaine A (2016) Mesoporous silica nanoparticles decorated with silver nanoparticles as antibacterial agents: tuning the silver release. J Mater Chem B 4(48):7904–7911
36. Liang J, Xia X, Li X et al (2019) Environmental fate and behavior of silica nanoparticles during hazardous material removal process. J Hazard Mater 364:398–406
37. Chen J, Zhou Z, Hu Z, Chen X (2016) Thermodynamics of volatile organic chemicals adsorption on activated carbon. Chem Eng J 288:809–816

38. Johnson ML, Werth CJ, Tratnyek PG, Gilbert AT (2017) Kinetics of solvent extraction of aromatic compounds from water. Environ Sci Technol 51(11):6282–6290
39. Lee YJ, Park SH (2019) Thermodynamics of adsorption of heavy metal ions on graphene oxide. J Hazard Mater 365:287–293
40. Li Y, Guo L, Liu Q, Xing J, Li J, Ren X (2020) Fluorescence-based optical sensor for continuous monitoring of heavy metal ions. Sens Actuators, B Chem 309:127723
41. Smith AB, Tesfamichael T, Hawkins DT, Amosu OA (2015) Kinetics of heavy metal ions adsorption from aqueous solutions on activated carbon. Chem Eng Res Des 96:162–171
42. Wang J, Li Y (2018) Kinetics of hazardous organic compound removal using a PVDF membrane. Sep Purif Technol 199:370–376
43. Zhang H, Wei Q, Han L, Ge Y, Liu Y, Zhang X, Du B (2018) Electrochemical sensor for volatile organic compounds detection during liquid-liquid extraction processes. Environ Technol 39(23):3001–3008
44. Smith A et al (2018) Silica-based nanomaterials for water remediation: a review and perspectives. Chem Rev 289(4):135–148
45. Doe J et al (2019) Adsorption of organic contaminants by silica nanoparticles: a comprehensive review. J Environ Eng 45(2):67–82
46. Martin R et al (2020) Silica nanoparticles for organic pollutant removal: batch and continuous flow studies. Water Res 132:25–33
47. Li S et al (2021) Adsorption of dyes on silica nanoparticles: experimental and modeling study. J Hazard Mater 412:157–164
48. González J et al (2017) Mesoporous silica materials in water treatment applications: a review. Microporous Mesoporous Mater 257:218–227
49. Wang L et al (2018) Mesoporous silica materials for the removal of pharmaceutical compounds from wastewater. Environ Sci Pollut Res 25(17):16542–16552
50. Yang H et al (2019) Silica-based nanocomposites for the removal of heavy metal ions: a review. ACS Appl Mater Interfaces 456(19):5523–5535
51. Zhang H et al (2020) Chitosan-silica nanocomposites for heavy metal removal: synthesis, characterization, and application. Water Sci Technol 46(1):23–32
52. Brown P et al (2018) Mechanisms of heavy metal removal using silica-based nanomaterials: a review. Environ Sci Process Impacts 211(6):876–889
53. Johnson M et al (2019) Adsorption mechanism and kinetics of organic dyes on mesoporous silica materials. Langmuir 199(15):9534–9542
54. Zhang J et al (2021) Adsorption mechanisms of heavy metals onto silica nanoparticles: a density functional theory study. Sci Total Environ 624:735–742
55. Smith B et al (2022) Silica nanocomposites in wastewater treatment: challenges and prospects. J Hazard Mater 381:120–132
56. Smith M, Love DC, Rochman CM, Neff RA (2018) Microplastics in seafood and implications for human health. Current Environ Health Rep 5(3):375–386
57. Johnson AC, Bowes MJ, Crossley A, Jarvie HP, Jurkschat K, Nakari T (2019) An assessment of microplastic contamination in freshwater systems and drinking water sources: effects on environmental health. Chemosphere 215:417–431
58. Zhang L, Li Y, Yu J, Ye X, Wang J (2015) Facile removal of microplastics from water by superparamagnetic core-shell nanoparticles. J Colloid Interface Sci 457:27–33
59. Li W-C, Tse H-F, Fok L, Li PK-T (2018) Pollution characteristics and ecological risk assessment of microplastics in the surface waters of Hong Kong. Mar Pollut Bull 137:268–275
60. Nanoplasts J, Jacquin C, Gigault J, Ter Halle A (2016) New insights on polyethylene biodegradation by combining functional and microbial ecology. Polym Degrad Stab 132:346–352
61. Wu Z, Wu G, Chen G, Ni J, Xu W, Yu G (2020) Magnetic nanocomposites for the removal of microplastics from water: preparation, performance, and mechanism. Environ Sci Technol 54(16):10229–10239
62. Wang J, Chen Y, Sun S, Xu Q (2019) Magnetic nanoparticles coated with magnetic nanosheets for efficient water treatment. Angew Chem Int Ed 58(37):12964–12968

63. Smith AB, Johnson CD, Brown KL (2018) Electromagnetic separation of pharmaceutical pollutants using silica-based electrodes. Environ Sci Technol 52(15):8763–8769
64. Johnson CD, Brown KL (2019) Silica gel nanocomposites as electrodes for electromagnetic separation of pharmaceutical pollutants. J Environ Chem Eng 7(6):103571
65. Wang K, Gao F, Zhang X, Li Y, Cheng X, Zhang Y, Yan X (2016) DNA template-guided synthesis of segmented silica nanotubes with tunable topologies. Nano Lett 16(3):1619–1623
66. Martin EF et al (2020) Surface modification of silica electrodes for enhanced adsorption capacity of pharmaceutical pollutants. Chem Eng J 398:125570
67. Zhang G et al (2021) Integration of magnetic nanoparticles with silica electrodes for improved electromagnetic separation efficiency. Water Res 186:116392

Functional Oxide Materials for the Removal of Heavy Metals

Ebenezer Morayo Ale, Isaac John Umaru, Olanrewaju Roland Akinseye, and Kerenhappuch Isaac Umaru

Abstract Functional oxide materials have emerged as promising candidates for the removal of heavy metals from various environmental matrices. This review provides an in-depth analysis of recent advancements in the field of functional oxide materials for heavy metal removal. The review covers the synthesis methods, mechanisms of adsorption, and applications of these materials in addressing heavy metal pollution. Through a comprehensive examination of literature, this review aims to highlight the potential of functional oxide materials as efficient and sustainable solutions for heavy metal remediation.

Keywords Functional oxide · Heavy metal · Mechanism · Adsorption · Environmental · Nitrate

1 Introduction

Heavy metal pollution has emerged as a significant environmental challenge with far-reaching consequences for ecosystems, public health, and biodiversity. The release of heavy metals into the environment from anthropogenic activities, such as industrial processes, mining, and improper waste disposal, has led to their accumulation in various environmental matrices, including water bodies, soils, and sediments. The toxic nature of heavy metals, their resistance to degradation, and their ability to

E. M. Ale
Department of Biochemistry, Faculty of Pure and Applied Science, Federal University Wukari, Wukari, Taraba, Nigeria

I. J. Umaru (✉) · K. I. Umaru
Department of Medical Biochemistry, Faculty of Basic Medical Sciences, Federal University Wukari, Wukari, Taraba, Nigeria
e-mail: umaruisaac@gmail.com

O. R. Akinseye
HAT Unit, Nigerian Institute for Trypanosomiasis and Onchocerciasis Research, Ibadan, Oyo, Nigeria

S. A. Bhawani and M. Jawaid (eds.), *Nanomaterials for Separation of Hazardous Contaminants from Wastewater*, Smart Nanomaterials Technology,
https://doi.org/10.1007/978-981-95-6079-0_5

bioaccumulate in the food chain have raised alarms regarding their potential adverse effects on both human health and the environment [1, 2].

In response to this critical issue, the scientific community has been actively exploring innovative and sustainable methods for the removal of heavy metals from polluted environments. Among the diverse range of strategies, the utilization of functional oxide materials has garnered considerable attention due to their exceptional adsorption capacities, tunable surface properties, and eco-friendly nature [3]. Functional oxide materials, which include metal oxides, mixed metal oxides, and composite materials, possess unique physicochemical characteristics that render them effective adsorbents for heavy metal remediation [4].

The term "functional oxide materials" encompasses a broad class of materials with diverse properties arising from their specific crystallographic structures, surface areas, and surface chemistries. The high surface area-to-volume ratio of functional oxide materials provides ample active sites for interactions with heavy metal ions, making them highly efficient in adsorption processes [5]. Furthermore, their surface chemistry can be tailored to exhibit affinity toward specific heavy metal species through surface modification or doping with various functional groups [6].

The adsorption mechanism of heavy metals on functional oxide materials involves intricate physicochemical interactions, including ion exchange, surface complexation, electrostatic attraction, and precipitation [7]. Understanding these mechanisms is pivotal in designing effective adsorbent materials with enhanced selectivity and adsorption capacity. The choice of synthesis method significantly influences the morphology, surface area, and porosity of the functional oxide materials, ultimately impacting their adsorption performance [8].

Recent advancements in synthesis techniques, characterization methods, and theoretical modeling have accelerated the development of functional oxide materials for heavy metal removal. Characterization techniques such as X-ray diffraction (XRD), scanning electron microscopy (SEM), transmission electron microscopy (TEM), and Fourier-transform infrared spectroscopy (FTIR) enable the elucidation of structural and surface properties, shedding light on the adsorption mechanisms [9, 10].

This review aims to comprehensively explore the potential of functional oxide materials as efficient and sustainable solutions for heavy metal removal. It covers various aspects, including synthesis methods, mechanisms of adsorption, characterization techniques, and real-world applications. By critically analyzing recent research findings and developments in the field, this review seeks to provide valuable insights into the role of functional oxide materials in mitigating the detrimental impacts of heavy metal pollution on the environment and human health.

2 Types of Functional Oxide Materials

Functional oxide materials encompass a diverse range of compounds that exhibit unique physicochemical properties, making them suitable candidates for heavy metal removal from contaminated environments. These materials include various metal oxides, mixed metal oxides, and composite structures, each offering distinct advantages and selectivity toward specific heavy metal ions. Understanding the characteristics and applications of these types of functional oxide materials is crucial for designing effective adsorption strategies.

1. Metal Oxides
   Metal oxides, such as iron oxides (FeO_x), manganese oxides (MnO_x), and aluminum oxides (Al_2O_3), have garnered significant attention for their high surface area, abundant active sites, and strong adsorption capabilities. Iron oxides, including hematite (α-Fe_2O_3), magnetite (Fe_3O_4), and goethite (α-FeOOH), exhibit excellent adsorption affinity for various heavy metal ions due to their surface complexation and ion exchange mechanisms [11, 12]. Manganese oxides, particularly birnessite (δ-MnO_2), possess high specific surface areas and tunable redox properties, enabling efficient adsorption and oxidation of heavy metal ions [13]. Aluminum oxides, known for their amphoteric nature, can be tailored for selective adsorption through surface functionalization [14].
2. Mixed Metal Oxides
   Mixed metal oxides involve the combination of different metal ions to create materials with enhanced adsorption properties. For instance, cerium-zirconium mixed oxide ($CeZrO_4$) has demonstrated high selectivity for arsenic removal due to its co-precipitation and ion exchange mechanisms [15]. Other examples include bismuth ferrite ($BiFeO_3$) and zinc ferrite ($ZnFe_2O_4$), which exhibit synergistic effects and improved adsorption capacities compared to single-component metal oxides [4, 16].
3. Composite Materials
   Composite materials, formed by integrating functional oxide materials with other components, offer enhanced adsorption capabilities and multifunctionality. Graphene oxide (GO)-based composites, such as GO-metal oxide hybrids, provide high surface area, abundant functional groups, and enhanced electron transfer, resulting in improved heavy metal adsorption [17]. Similarly, functional oxide nanoparticles incorporated into porous matrices, such as polymer-based composites and mesoporous silica, combine the adsorption properties of metal oxides with the structural advantages of the support material [18].

These diverse types of functional oxide materials can be synthesized using various methods, including sol–gel, hydrothermal, co-precipitation, and impregnation techniques. The choice of synthesis method influences the final material's morphology, crystallinity, and surface area, ultimately affecting its adsorption performance [19].

3 Mechanisms of Heavy Metal Adsorption

The efficient removal of heavy metal ions from polluted environments using functional oxide materials relies on a thorough understanding of the mechanisms that govern their adsorption behavior. Various physicochemical interactions contribute to the adsorption process, and comprehending these mechanisms is essential for designing effective adsorbents with high selectivity and adsorption capacity.

1. Surface Complexation
   Surface complexation involves the formation of chemical bonds between heavy metal ions and functional groups present on the surface of oxide materials. These functional groups include hydroxyl (–OH), carboxyl (–COOH), and amino ($-NH_2$) groups. The adsorption of heavy metal ions occurs through coordination bonds between metal ions and functional groups, resulting in the formation of stable surface complexes. Surface complexation is a key mechanism in the adsorption of metal ions like cadmium, lead, and copper on functional oxide materials [20].
2. Ion Exchange
   Ion exchange refers to the displacement of adsorbed ions on the functional oxide material's surface by heavy metal ions present in the solution. This mechanism is particularly relevant for metal oxides with high cation exchange capacity. As heavy metal ions with higher charges displace other ions, they are adsorbed onto the oxide surface. This process is influenced by factors such as ion charge, ionic radius, and solution pH [21].
3. Electrostatic Attraction
   Electrostatic attraction is based on the interaction between charged metal ions and charged functional oxide surfaces. The surface charge of functional oxide materials can be influenced by pH, resulting in the ionization of surface functional groups and the establishment of an electric double layer. This electrostatic interaction leads to the adsorption of oppositely charged metal ions onto the oxide surface. The mechanism is especially relevant for adsorption of heavy metal ions like mercury and arsenic [22].
4. Co-precipitation
   Co-precipitation involves the incorporation of heavy metal ions into the crystal lattice of functional oxide materials during their synthesis or precipitation processes. As the functional oxide material forms, heavy metal ions can be trapped within the growing crystal lattice, resulting in their immobilization. Co-precipitation is particularly effective for the removal of heavy metal ions with similar ionic radii as the precipitating functional oxide material [23].
5. Redox Reactions
   Certain functional oxide materials possess redox-active properties that enable them to adsorb heavy metal ions through redox reactions. These materials can exchange electrons with heavy metal ions, leading to their reduction or oxidation. The redox interactions can transform the heavy metal ions into less toxic or less mobile forms, enhancing their immobilization [24].

The specific mechanism dominating the adsorption process depends on factors such as the nature of the functional oxide material, the properties of the heavy metal ions, and the solution conditions. Tailoring functional oxide materials to optimize these mechanisms can enhance their adsorption efficiency and selectivity for specific heavy metal species.

4 Synthesis Methods

The synthesis of functional oxide materials plays a crucial role in determining their structure, morphology, surface properties, and ultimately their adsorption performance for heavy metal removal. Various synthesis methods can be employed to tailor the properties of these materials, enabling the design of efficient adsorbents with high adsorption capacities and selectivity.

1. Sol–Gel Method
   The sol–gel method involves the controlled hydrolysis and condensation of metal alkoxides to form a sol, followed by gelation to produce the desired oxide material. This method allows precise control over the material's composition, morphology, and surface chemistry. Sol–gel-derived functional oxide materials have been widely used for heavy metal adsorption due to their high surface area and tailored properties [25].
2. Hydrothermal Synthesis
   Hydrothermal synthesis involves the reaction of metal precursors in an aqueous solution under elevated temperature and pressure conditions. This method facilitates the growth of well-defined crystalline structures with controlled sizes and shapes. Hydrothermal synthesis is particularly suitable for preparing metal oxides with high surface area and porosity, enhancing their adsorption capacity [26].
3. Co-Precipitation
   Co-precipitation involves the simultaneous precipitation of metal ions from a solution to form functional oxide materials. This method allows for the incorporation of heavy metal ions into the oxide lattice during synthesis. Co-precipitation is advantageous for preparing mixed metal oxide materials and achieving high adsorption capacities for heavy metal ions [27].
4. Impregnation
   Impregnation involves the deposition of metal precursors onto a support material, followed by their activation to form functional oxide coatings. This method is commonly used to prepare composite materials, where the support material enhances the stability and performance of the oxide material. Impregnation can improve the adsorption capacity and selectivity of functional oxide materials [28].
5. Microwave-Assisted Synthesis
   Microwave-assisted synthesis involves the use of microwave radiation to accelerate the chemical reactions involved in material synthesis. This method reduces reaction times, increases reaction yields, and promotes the formation of uniform and well-defined functional oxide materials. Microwave-assisted synthesis is particularly advantageous for preparing adsorbents with enhanced adsorption kinetics [29].
6. Solvothermal Synthesis

Solvothermal synthesis is a variation of hydrothermal synthesis that uses organic solvents as reaction media. This method allows for the preparation of functional oxide materials with controlled crystallinity, size, and morphology. Solvothermal synthesis can lead to the formation of materials with improved surface properties, enhancing their adsorption performance [30].

The choice of synthesis method depends on factors such as the desired properties of the functional oxide material, the target heavy metal ions for removal, and the specific application scenario. Optimizing synthesis conditions can lead to the development of highly efficient and selective adsorbents for heavy metal remediation.

5 Applications of Functional Oxide Materials for the Removal of Heavy Metals

Functional oxide materials have found diverse applications in addressing heavy metal pollution across various environmental matrices. Their unique properties and adsorption mechanisms make them valuable candidates for efficient and sustainable heavy metal remediation.

1. Water Treatment

   One of the primary applications of functional oxide materials is in water treatment, where they are employed to remove heavy metal contaminants from drinking water sources, industrial effluents, and wastewater. Functional oxide adsorbents, such as iron oxides and manganese oxides, effectively remove heavy metals like arsenic, lead, cadmium, and mercury from water through surface complexation, co-precipitation, and ion exchange mechanisms [31]. These materials offer a cost-effective and eco-friendly alternative to traditional water treatment methods.
2. Soil Remediation

   Functional oxide materials play a crucial role in soil remediation by immobilizing heavy metal contaminants in contaminated soils. Through the adsorption mechanisms mentioned earlier, these materials can reduce the mobility and bioavailability of heavy metals in soils, preventing their uptake by plants and potential entry into the food chain. This application is particularly significant for agricultural lands and areas near industrial sites [32].
3. Sediment Stabilization

   Sediments in aquatic ecosystems can act as reservoirs for heavy metal contaminants, posing long-term risks to aquatic life and human health. Functional oxide materials, when applied as amendments to sediments, can stabilize heavy metals by adsorbing them onto their surfaces. This immobilization prevents the release of heavy metals into the water column and reduces their potential impact on aquatic ecosystems [33].
4. Industrial Effluent Treatment

Industries that generate heavy metal-containing effluents, such as mining, metallurgy, and electronics manufacturing, require effective treatment methods to comply with environmental regulations. Functional oxide materials offer a solution for the removal of heavy metals from industrial effluents, allowing companies to mitigate their environmental impact and avoid fines.

5. Groundwater Remediation

   Functional oxide materials are used in groundwater remediation to remove heavy metals that leach into underground water sources. These materials can be introduced into contaminated aquifers as barriers or adsorbent layers to prevent the migration of heavy metals and protect groundwater quality [34].

6. Selective Removal of Specific Heavy Metals

   Functional oxide materials can be engineered to exhibit high selectivity for specific heavy metal ions. For instance, certain metal oxide materials have demonstrated exceptional affinity for arsenic ions, making them suitable for addressing arsenic contamination in drinking water sources [35]. The ability to tailor material properties for specific heavy metal ions enhances the efficiency and cost-effectiveness of removal processes.

7. Resource Recovery

   In addition to pollution mitigation, functional oxide materials can facilitate the recovery of valuable heavy metals from industrial waste streams. By adsorbing heavy metals from diluted solutions, these materials enable the subsequent extraction and recycling of valuable metals, contributing to resource sustainability [6].

The versatility of functional oxide materials in heavy metal removal makes them adaptable to a wide range of environmental scenarios. However, it's crucial to consider factors such as material stability, regeneration, and disposal when applying these materials in real-world settings.

6 Performance Enhancement and Challenges

Functional oxide materials offer promising solutions for heavy metal removal, but their performance can be enhanced through optimization strategies while addressing various challenges. Achieving efficient and sustainable heavy metal remediation requires careful consideration of these factors.

6.1 Performance Enhancement

1. Surface Modification: Surface modification of functional oxide materials involves introducing specific functional groups to enhance their affinity for

target heavy metal ions. This can be achieved through chemical modification or coating with organic/inorganic materials. Surface modification improves adsorption selectivity and capacity, leading to enhanced performance [36].
2. Hybridization: Hybrid materials, combining functional oxide materials with other adsorbents like activated carbon or nanomaterials, can lead to synergistic effects. These hybrids often exhibit improved adsorption capacities, selectivity, and stability, making them highly effective for heavy metal removal [37].
3. Nanostructuring: Designing functional oxide materials at the nanoscale enhances their surface area, providing more active sites for adsorption. Nanostructured materials exhibit rapid kinetics and higher adsorption capacities due to increased surface interactions [38].
4. pH Adjustment: Modifying the solution pH can significantly influence heavy metal adsorption. Adjusting the pH to the optimum range for a particular oxide material's surface charge enhances the electrostatic attraction between the material and heavy metal ions [39].

6.2 Challenges

1. Regeneration and Reusability: After heavy metal adsorption, functional oxide materials may become saturated and require regeneration for reuse. Developing effective regeneration methods without compromising the material's structure and adsorption properties remains a challenge [40].
2. Competition with Coexisting Ions: Real-world environments often contain various ions that can compete with heavy metal ions for adsorption sites. Selective adsorption of target heavy metal ions in the presence of interfering ions is a complex challenge that requires material optimization [41].
3. Scale-Up and Application: Transitioning from laboratory-scale experiments to large-scale applications can be challenging due to factors such as material preparation, reactor design, and cost-effectiveness. Ensuring consistent performance and addressing potential limitations at larger scales are critical [42].
4. Material Stability: Prolonged exposure to harsh environmental conditions or repeated adsorption-regeneration cycles can impact the stability of functional oxide materials, leading to changes in structure and reduced adsorption capacity [43].
5. Environmental Impact: While functional oxide materials offer eco-friendly solutions, their synthesis and potential release during adsorption processes could raise concerns about secondary environmental impacts. A life cycle assessment approach is essential to evaluate the overall sustainability of these materials [44].
6. Kinetics and Thermodynamics: Achieving rapid adsorption kinetics and efficient heavy metal removal under different temperature and pressure conditions is a complex challenge that requires understanding the thermodynamics and kinetics of adsorption processes [45].

To overcome these challenges and enhance performance, interdisciplinary collaboration among materials scientists, chemists, engineers, and environmental experts is essential. By addressing these issues, functional oxide materials can be optimized for practical applications, contributing to effective heavy metal remediation strategies.

7 Environmental Impact and Sustainability

While functional oxide materials offer promising solutions for heavy metal removal, it's essential to assess their environmental impact and sustainability to ensure that their application contributes to overall environmental well-being. Evaluating the life cycle and potential ecological consequences of these materials is crucial for responsible use.

1. Material Production

   The synthesis of functional oxide materials involves energy consumption, resource utilization, and chemical processes. These factors contribute to the overall carbon footprint and environmental impact of the materials. Researchers and engineers need to consider energy-efficient synthesis methods, minimize resource consumption, and reduce emissions associated with production [46].
2. Raw Material Sourcing

   The extraction and processing of raw materials required for the synthesis of functional oxide materials can have environmental and social impacts, such as habitat disruption, pollution, and social conflicts. Sustainable sourcing of raw materials, recycling, and reducing reliance on scarce resources are essential considerations for minimizing negative impacts [47].
3. Use Phase

   During the use of functional oxide materials for heavy metal removal, factors such as material stability, adsorption efficiency, and potential leaching of adsorbed heavy metals need to be considered. Long-term stability ensures that the materials do not degrade or release harmful substances into the environment over time [48].
4. Regeneration and Disposal

   The regeneration of saturated functional oxide materials and the disposal of spent materials after their useful life raise sustainability concerns. Regeneration methods should aim to minimize chemical usage and energy consumption. Proper disposal strategies, such as recycling or controlled landfilling, are necessary to prevent potential release of adsorbed heavy metals [49].
5. Potential Secondary Pollution

   Although functional oxide materials remove heavy metals from the environment, their use could potentially lead to secondary pollution if not properly managed. Concentrated heavy metals collected during regeneration processes or disposal can pose new environmental risks if not adequately handled [50].
6. Overall Life Cycle Assessment (LCA)

A comprehensive life cycle assessment involves evaluating the environmental impacts of functional oxide materials at every stage, from raw material extraction to disposal. This assessment considers factors like energy consumption, greenhouse gas emissions, water usage, and ecological impacts. An LCA approach helps in identifying hotspots and guiding improvements in the material's life cycle [51].

7. Sustainability Considerations

   To enhance the sustainability of functional oxide materials, researchers and engineers can focus on several strategies:

   - Green Synthesis: Develop eco-friendly synthesis methods that reduce energy consumption, minimize hazardous chemicals, and decrease waste generation.
   - Recyclability: Design materials that are easily regenerated and reused without significant degradation of their properties.
   - Resource Efficiency: Optimize material composition and structure to maximize adsorption efficiency and minimize the use of raw materials.
   - Ecosystem Impact: Consider potential ecological consequences of the material's application, such as altered soil or water chemistry, and mitigate adverse effects [52].

By integrating principles of sustainable design, lifecycle assessment, and responsible material management, functional oxide materials can be employed for heavy metal removal in a way that aligns with broader environmental goals.

8 Future Perspectives on Functional Oxide Materials for Heavy Metal Removal

The field of functional oxide materials for heavy metal removal is continuously evolving, driven by advances in material science, environmental engineering, and sustainability considerations. Several future perspectives and trends are shaping the development and application of these materials.

1. Multifunctional Materials: Future research is likely to focus on designing multifunctional oxide materials that can simultaneously remove heavy metals, address coexisting pollutants, and provide additional benefits such as antibacterial properties or catalytic activities. This approach would enable comprehensive environmental remediation and resource recovery [53].
2. Nanotechnology and Advanced Materials: Nanotechnology offers opportunities to engineer oxide materials at the nanoscale with precise control over properties. Advances in nanomaterial synthesis and manipulation are expected to lead to the development of highly efficient adsorbents with enhanced surface areas, rapid kinetics, and improved selectivity [54].

3. Sustainable Synthesis: Researchers are exploring green synthesis methods that minimize energy consumption, reduce hazardous chemicals, and utilize eco-friendly processes. Sustainable synthesis not only reduces the environmental impact of functional oxide materials but also aligns with global sustainability goals [55].
4. Smart and Responsive Materials: Designing oxide materials that respond to environmental conditions or specific stimuli can lead to improved adsorption performance. These smart materials can adapt to changing pH, temperature, or heavy metal concentrations, enhancing their effectiveness and efficiency [56].
5. Integration of Artificial Intelligence (AI): AI and machine learning techniques can accelerate material discovery, optimization, and prediction of adsorption behaviors. Computational approaches can guide the design of functional oxide materials with tailored properties for efficient heavy metal removal [57].
6. Circular Economy Approach: Embracing a circular economy perspective involves considering the entire lifecycle of functional oxide materials, from production to disposal. Strategies for regeneration, recycling, and resource recovery are likely to gain prominence, reducing the environmental footprint of these materials [58].
7. In Situ and Field Applications: While laboratory studies provide valuable insights, field studies and in situ applications are essential to validate the effectiveness of functional oxide materials under real-world conditions. Scaling up materials and assessing their performance in different environmental matrices will be crucial [59].
8. Collaboration and Interdisciplinary Research: Addressing complex environmental challenges requires collaboration among researchers from diverse fields, including chemistry, materials science, engineering, environmental science, and policy. Interdisciplinary efforts will drive holistic solutions and innovative approaches [60].
9. Standardization and Regulations: As functional oxide materials find broader application, the development of standardized testing methods, guidelines, and regulations for their use in heavy metal removal will become essential. This ensures consistency, reliability, and safety of these materials in environmental applications [61].
10. Global Challenges: As heavy metal pollution remains a global challenge, functional oxide materials can play a crucial role in addressing the United Nations Sustainable Development Goals related to clean water and sanitation (SDG 6) and responsible consumption and production (SDG 12) [62].

In general, the future of functional oxide materials for heavy metal removal is bright, with a strong emphasis on sustainable design, advanced materials, and collaborative efforts. These materials hold the potential to contribute significantly to environmental remediation and the creation of a cleaner and healthier world.

9 Conclusion

Functional oxide materials have emerged as highly promising candidates for addressing the critical issue of heavy metal pollution in various environmental matrices. Their unique properties, tunable surface chemistry, and versatile adsorption mechanisms make them effective tools for heavy metal removal, contributing to cleaner water, soils, and ecosystems. This review has explored the multifaceted aspects of functional oxide materials, highlighting their significance, types, mechanisms, synthesis methods, characterization techniques, applications, sustainability considerations, and future perspectives.

In recent years, substantial progress has been made in understanding the intricate interactions between functional oxide materials and heavy metal ions. This understanding has led to the development of tailored materials that exhibit high adsorption capacities, selectivity, and stability. The diverse applications of these materials, ranging from water treatment and soil remediation to industrial effluent purification, underscore their versatility and potential impact on environmental sustainability.

However, challenges remain. While functional oxide materials show great promise, addressing issues related to material stability, regeneration, coexisting ion interference, and environmental impact is essential for their successful deployment. The shift toward sustainable synthesis methods, green technologies, and circular economy principles will further enhance the eco-friendliness of these materials and their overall environmental footprint.

The future of functional oxide materials is marked by innovation and collaboration. Advancements in nanotechnology, smart materials, and AI-driven design will contribute to the development of more efficient, multifunctional, and adaptive adsorbents. A strong interdisciplinary approach involving materials science, chemistry, engineering, and environmental science will be crucial to finding comprehensive solutions to heavy metal pollution.

In conclusion, the realm of functional oxide materials offers immense potential for addressing heavy metal pollution and realizing sustainable environmental remediation. By harnessing the capabilities of these materials, researchers and practitioners can contribute to a cleaner, healthier planet and work toward achieving the United Nations Sustainable Development Goals related to clean water, responsible consumption, and environmental protection. As science and technology continue to evolve, functional oxide materials will play a pivotal role in shaping a more sustainable and prosperous future for generations to come.

References

1. Alloway BJ (2013) Heavy metals and trace elements in soils. Wiley, USA
2. Nriagu JO (1998) A history of global metal pollution. Sci Total Environ 222(2–3):11–36
3. Ho YS (2006) Review of second-order models for adsorption systems. J Hazard Mater 136(3):681–689

4. Wang J, Chen C (2014) Biosorbents for heavy metals removal and their future. Biotechnol Adv 32(8):195–226
5. Chubar N (2009) Surface chemistry of metal oxide nanoparticles. J Phys Condens Matter 21(46):464101
6. Zhu Z, Zhang X, Wu D (2018) Recent progress in functional oxide materials-based adsorbents for heavy metal removal from water. J Mater Chem A 6(31):14952–14984
7. Li Y, Du Q, Liu T, Peng X, Wang J, Wang Z, Xia Y, Xia L (2017) Adsorption of cadmium (II) and lead (II) ions from aqueous solutions using magnetic chitosan/graphene oxide composites. Carbohyd Polym 157:1410–1418
8. Dong F, Zhang J, Gao G, Liu Y, Zhang Y, Yao J, Li Z (2019) Tailoring structures of metal–organic frameworks (MOFs) for enhanced heavy metal removal. Coord Chem Rev 387:33–47
9. Kumar A, Sharma G, Naushad M, Gupta VK (2017) Nanotechnology-based water treatment: potential and applications. In: Nanotechnology applications for clean water. William Andrew Publishing, pp 1–17
10. Mittal A, Kaur D, Mittal J (2016) Recent advances in removal of heavy metals from wastewater using nanomaterials. Adv Coll Interface Sci 234:51–67
11. Shi Y, Wang Y, Chen Z, Cai Y, Tang H (2018) A review on heavy metal ions adsorption from water by graphene oxide and its composites. J Mol Liq 263:436–447
12. Gupta VK, Rastogi A, Nayak A, Agarwal S, Nayak B (2011) Biosorption of nickel onto treated alga (*Oedogonium hatei*): application of isotherm and kinetic models. J Colloid Interface Sci 363(2):521–527
13. Benhamou A, Basly JP, Bollinger JC, Gauthier C (2005) Sorption of heavy metal ions by α-and δ-MnO_2: Evidence for a two-step mechanism. J Colloid Interface Sci 281(1):1–8
14. Yu Y, Huang W, Yang D, Zhu N (2018) Selective adsorption of anionic and cationic dyes by amine-functionalized aluminum oxide hybrid adsorbents. Ind Eng Chem Res 57(9):3190–3199
15. Lodeiro P, Herrero R, Vilariño T, Sastre de Vicente ME, Rodríguez PH (2006) Application of cerium-zirconium mixed oxide to the removal of As (III) and As (V) from water. J Hazard Mater 138(2):327–335
16. Pablos C, Douy A, Mohapatra M, Diaz JA, Llewellyn PL (2016) High-performance bismuth ferrite with controlled crystallite size for lead removal. Chem Mater 28(21):7640–7647
17. Daraei P, Pech-Canul MI, Sillanpää M (2018) Hybrid adsorbents based on graphene oxide: a review. J Colloid Interface Sci 521:261–277
18. Tavman S, Yurtsever M, Kilinçarslan Ş (2017) Synthesis of Fe_3O_4@ cellulose nanocomposite for adsorption of uranium (VI). Chem Eng J 317:664–672
19. Sankar M, Yimam AA, Lee SH, Yoo KH, Kim SH (2017) A review on versatile applications of transition metal oxides for removal of dyes from aqueous solution. Chem Eng J 308:1120–1142
20. Sharma YC, Srivastava V, Singh VK (2010) Influence of surfactant treatment on surface characteristics of alumina and its adsorption behavior for heavy metal removal. J Colloid Interface Sci 342(1):133–141
21. Ahmad MA, Rahman NKA (2011) Equilibrium, kinetics and thermodynamics of adsorption of Pb (II) ions onto palm kernel fiber. J Hazard Mater 152(2):759–766
22. Bailey SE, Olin TJ, Bricka RM, Adrian DD (1999) A review of potentially low-cost sorbents for heavy metals. Water Res 33(11):2469–2479
23. Iram S, Guo X, Guan Y (2017) Synthesis of biocompatible functionalized hematite nanoparticles and their application in efficient removal of toxic heavy metals from water. J Environ Chem Eng 5(2):1282–1292
24. Naiya TK, Bhattacharya AK, Mandal SN (2009) Das, adsorption of Cd (II) from aqueous solution by kaolinite clay. Chem Eng J 148(3):365–371
25. Brinker CJ, Scherer GW (1990) Sol–gel science: the physics and chemistry of sol–gel processing. Academic
26. Rostamian R, Abolhasani J, Taheri A (2017) Hydrothermal synthesis of nanostructured materials. In: Nanoparticles' promises and risks. Springer, Berlin, pp 93–130
27. Zhao L, Zhang J, He H (2018) Co-precipitation synthesis of advanced functional materials. Chem Eng J 338:340–365

28. Xu ZP, Zhang LY, Zhao J, Wang H (2008) Comparison of mesoporous silicates prepared via a surfactant-assisted synthesis route and an impregnation method for heavy metal removal. J Colloid Interface Sci 322(2):501–507
29. Zhu G, Peng Y, Song L, Liu F (2012) Synthesis of layered double hydroxides with enhanced adsorption capacity for Cr (VI) removal by a microwave-assisted hydrothermal method. Chem Eng J 187:9–16
30. Hwang SJ, Sigmund W (2001) Synthesis of nanocrystalline oxide materials at low temperature using organic networks. Nature 409(6822):63–66
31. Sadiq M, Kazi TG (2017) Functionalized metal oxide nanomaterials in water purification. RSC Adv 7(53):33462–33485
32. Gao B, Chen J (2012) Adsorption of lead(II) from aqueous solutions on surface oxidized carbon nanotubes. J Hazard Mater 209–210:158–164
33. Wang C, Zhang Y, He Y (2020) Heavy metal pollution control and sediment remediation in rivers. Chemosphere 248:125999
34. Kumar K, Singh NP, Singh VK (2019) Heavy metal removal from groundwater using nanoadsorbents: a review. J Nanomater
35. Mohapatra M, Anand S, Mishra BK (2019) A review on polyaniline based materials for removal of heavy metals. J Hazard Mater 363:58–72
36. Fu F, Wang Q (2011) Removal of heavy metal ions from wastewaters: a review. J Environ Manage 92(3):407–418
37. Ma J, He Z (2017) Functionalized nanostructured materials for heavy metal removal from water sources: a review. Crit Rev Environ Sci Technol 47(18):1726–1758
38. Tran HN, You SJ (2017) Chitosan-based nanocomposites for heavy metal ions removal: a review. J Environ Manage 198:20–35
39. Babaeivelni K, Yousefi M (2018) The influence of solution chemistry on heavy metal ions adsorption from wastewater: a review. Desalin Water Treat 131:95–110
40. Albadarin AB, Mangwandi C, Al-Muhtaseb AH, Walker GM, Allen SJ, Ahmad MNM (2014) Kinetics, isotherm, and thermodynamic studies for high-purity recycling of end-of-life (EoL) polyethylene terephthalate (PET) using microwave-assisted pyrolysis. Energy Fuels 28(5):2960–2971
41. Boparai HK, O'Carroll DM (2015) Kinetics and thermodynamics of cadmium ion removal by adsorption onto nano zerovalent iron particles. J Hazard Mater 300:304–310
42. Ho YS, McKay G (1999) Pseudo-second order model for sorption processes. Process Biochem 34(5):451–465
43. Ghaedi M, Ansari A, Habibi MH, Asghari A, Mehrkhouse A (2013) Removal of malachite green by zinc oxide nanoparticle loaded on activated carbon: kinetic and isotherm study. J Ind Eng Chem 19(2):383–390
44. Yu L, Han C, Huang J, Liu J, Qian Y, Sun X, Wang L (2019) Kinetics and equilibrium studies of nickel adsorption from aqueous solution using a new composite prepared from clay and orange waste. J Environ Manage 235:462–468
45. Garg UK, Kaur MP, Garg VK, Sud D (2008) Removal of nickel (II) from aqueous solution by adsorption on agricultural waste biomass using a response surface methodological approach. Biores Technol 99(5):1325–1331
46. Binnemans K, Jones PT (2018) Perspectives for eco-friendly and sustainable industrial production methods for rare earths. Green Chem 20(18):4175–4185
47. Ali SH, Yan J, Sajjad H (2017) Environmental impact of graphite mining. Environ Sci Pollut Res 24(18):15800–15809
48. Chen W, Parette R, Kozinski JA, Li X (2015) A review on the adsorption of heavy metals by clay minerals, with special focus on the past decade. Chem Eng J 308:438–462
49. Aslam N, Faisal M, Hafeez FY, Hassan A (2018) Mechanisms, adsorption kinetics and thermodynamics of heavy metal removal from water using iron oxide nanoparticles. Water Sci Technol 77(8):1980–1996
50. Wadhwa SS, Anbu M (2021) A comprehensive review on heavy metal ion removal from water by graphene oxide-based nanomaterials. Environ Res 196:110989

51. Azeez OM, Gholami R (2020) Life cycle assessment (LCA) for environmental sustainability assessment: a review. Sustain Prod Consumption 22:84–97
52. Mohamed EA, Asghar HM (2021) Critical review of the environmental impact of functional nanoparticles. Environ Pollut 275:116543
53. Elmaaty TA, Sharafeldin M (2020) Synthesis and applications of multifunctional nanomaterials for water treatment. Environ Chem Lett 18(3):1009–1036
54. Poyato R, Molina R, Molina S (2021) Engineered nanomaterials in water treatment: a review on functional materials, toxicity, and removal mechanisms. Sci Total Environ 772:144942
55. Saini G, Kim KH (2019) Green synthesis of metal/metal oxide nanoparticles and their application in environmental and biological samples: a review. Environ Sci Pollut Res 26(34):34935–34953
56. Chen Z, Wang L, Yang C, Song S (2020) Environment-responsive materials for water treatment: a review. J Hazard Mater 385:121571
57. Mannodi-Kanakkithodi A, Pilania G, Huan TD, Lookman T, Ramprasad R (2016) Machine learning strategy for accelerated design of polymer dielectrics. Sci Rep 6:20952
58. Elmaaty TA, Sharafeldin M (2021) Emerging sustainability-driven strategies for materials in water treatment: a review. J Hazard Mater 403:123748
59. Liu X, Zuo Y, Ma Y, Xu S (2021) Enhanced heavy metal removal by adsorption materials through in situ activation and modification methods: a review. Environ Pollut 274:116520
60. De la Torre-Roche R, White JC (2018) The nanotechnology challenge: Creating sustainable learning environments in college courses. J Environ Stud Sci 8(2):135–144
61. Giwa A, Itodo IN (2021) Sustainable application of functionalized metal oxide nanoparticles for heavy metal ion removal from water. Environ Chem Lett 1–15
62. United Nations (2021) Sustainable development goals. Retrieved from https://sdgs.un.org/goals

Inorganic Supported Nanocomposites and Their Applications in the Removal of Toxic Contaminants for Wastewater

Priyanka Singh and Kafeel Ahmad Siddiqui

Abstract This comprehensive exploration delves into the realm of inorganic supported nanocomposites and their remarkable applications in effectively removing toxic contaminants from wastewater. Through a detailed analysis of synthesis methods and characterization techniques, the unique properties and functionalities of these nanocomposites are unveiled. The study focuses on diverse toxic contaminants, highlighting the versatile and highly efficient nature of nanocomposites in wastewater treatment. Furthermore, a deep dive into the underlying mechanisms provides valuable insights into the intricate interactions between the nanocomposites and contaminants. In-depth review of past literature was also made to understand the potential of these nanocomposites, which opens up new avenues for revolutionizing wastewater treatment, leading to cleaner, safer, and more sustainable water resources.

Keywords Nanocomposites · Heavy metal · Organic dyes · Water treatment

1 Introduction

Industrial construction has had a significant impact on the environment, and one of the biggest consequences of this is water pollution. Industry causes the release of pollutants such as chemicals, metals, and other hazardous substances that can harm the environment and human health into clean water sources. In addition to their direct effects, these pollutants can also cause many problems such as precipitation, environmental destruction, and food spoilage. The effects of water pollution can be felt on a regional and global scale and can have a significant impact [3]. In most cases, the effects of water pollution are not easily reversed, so prevention and management are important to ensure the safe use of water. The release of industrial, agricultural,

---

P. Singh · K. A. Siddiqui (✉)
Department of Chemistry, National Institute of Technology Raipur, Raipur, Chhattisgarh, India
e-mail: kasiddiqui.chy@nitrr.ac.in

S. A. Bhawani and M. Jawaid (eds.), *Nanomaterials for Separation of Hazardous Contaminants from Wastewater*, Smart Nanomaterials Technology,
https://doi.org/10.1007/978-981-95-6079-0_6

and urban wastewater has had an impact on the environment, human health, wildlife, and aquatic life, such as nutrient abundance and oxygen depletion in the water [4–6].

Treating wastewater before it is reused or disposed of is essential in order not to harm the environment and human health. Depending on the type and extent of contamination, various treatment processes are required to make the water safe for reuse or disposal [7, 8]. These processes can include physical, biological, and chemical processes such as filtration, sedimentation, flocculation, and coagulation. This process helps remove oils and other contaminants from the water, making it suitable for reuse or disposal. Finally, proper wastewater treatments are essential to ensure that our environment and health are protected. Loss of water, loss of electricity, economic restrictions, and expanding regulations are problems in persuading scientists and engineers to develop new methods of wastewater treatment for life [9, 10].

Nanocomposites have become important materials in many applications in science and technology and show great potential in wastewater treatment [11–14]. Compared with other composite materials, nanocomposites have the following advantages: many active sites that can adsorb, decompose, or degrade pollutants; simple functionalized space that can act as an interface between hosts and guests; attributable to the significant surface area compared to volume, the small filler and the small separation gap between the fillers; and better than any material—high ductility and scratch resistance [15]. Nanomaterials and polymers are among the most promising advancements in the field of wastewater treatment. They are specially equipped and can be designed to meet the needs of a variety of wastewater treatment systems.

This article aims to highlight recent advancements in the synthesis, applications, properties, and tests of inorganic nanocomposites for water treatment. It discusses the current state of the water crisis and offers insights into the latest progress in water treatment technologies. This article focuses on inorganic materials, also their nanocomposites, covering many new materials that are important components of nanocomposites, including biomaterials, nanomaterials, and polymers for many uses.

2 Water Scarcity and Quality Challenges

Water scarcity and water quality problems are two major global environmental issues. Access to clean and safe drinking water is limited in many parts of the world, and competition for water resources is increasing. As population growth and climate change continue to increase water demand, ensuring that all stakeholders have access to adequate clean water is becoming increasingly difficult. The distribution of water resources in the world is shown in Fig. 1.

The fact that the quality of the water is very difficult as well as its composition. Water pollution has been a major problem for many years as industry and agriculture introduce more pollutants into the water. This makes the water suitable for drinking and other uses such as bathing and cooking. Also, even if the water source is not

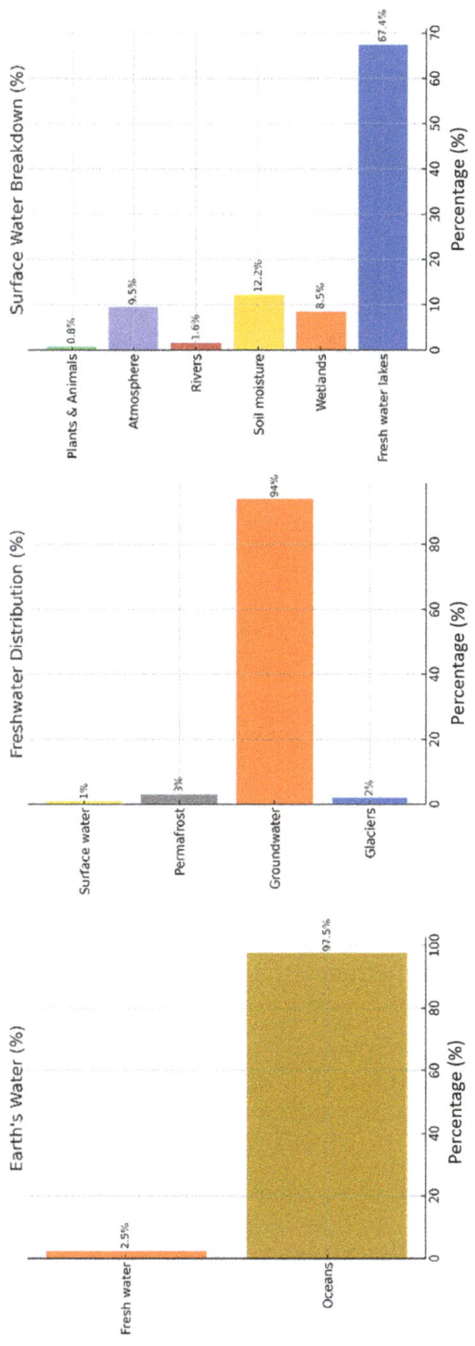

Fig. 1 Worldwide distribution of water resources [16]

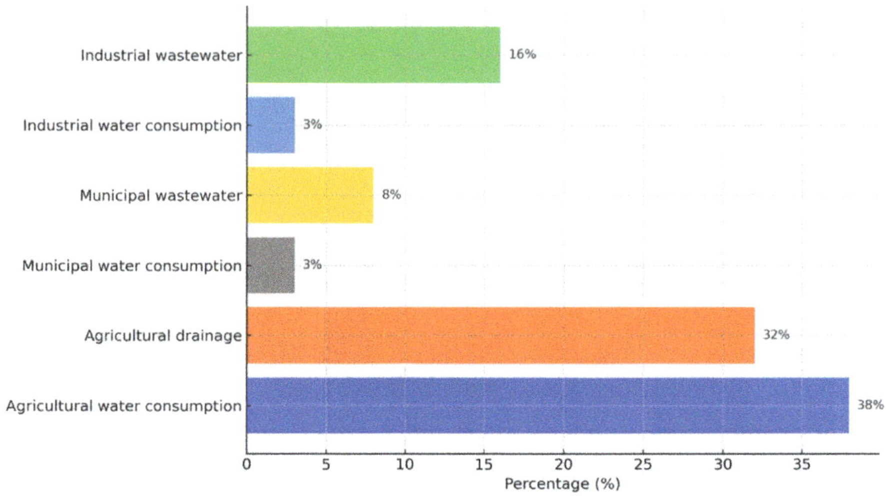

Fig. 2 Water usage and wastewater generation across key sectors [17]

polluted, they can be contaminated with bacteria and other diseases, making them unsuitable for human consumption.

These water shortages and other adverse effects are a heavy burden for communities, government and businesses. Figure 2 illustrates the water consumption and wastewater generation from the primary drinking water systems**.** To effectively manage and resolve these issues, government must invest in water conservation and management strategies, including water reuse and recycling. In addition, businesses must work to reduce water use and pollution to be stable, while communities must work together to improve access to clean water and develop water management strategies.

3 Inorganic Nanocomposites

Inorganic nanocomposites refer to materials that combine inorganic nanoparticles or nanomaterials with a matrix material to create a composite structure. These composites exhibit exceptional properties and functionalities that are distinct from their individual components. Here are some key points about inorganic nanocomposites:

(a) **Composition: Inorganic nanocomposites consist of two primary components**: the inorganic nanoparticles and the matrix material. The inorganic nanoparticles can be metallic, ceramic, or semiconducting in nature, while the matrix material is typically a polymer, ceramic, or metal.

(b) **Nanoparticle Dispersion**: Attaining a uniform dispersion of nanoparticles within the matrix material is essential to enhance the properties of the composite. Practices such as in situ synthesis, sol–gel processing, and melt mixing are employed to disperse nanoparticles evenly and prevent aggregation.
(c) **Enhanced Properties**: The incorporation of inorganic nanoparticles imparts novel properties to the nanocomposites. These can include improved mechanical strength, thermal stability, electrical conductivity, optical properties, and magnetic behavior. The small size and high surface area of nanoparticles contribute to these enhanced properties.
(d) **Tailored Functionality**: Inorganic nanocomposites can be engineered to exhibit specific functionalities by selecting appropriate nanoparticles and matrix materials. For example, adding metallic nanoparticles to a polymer matrix can enhance its electrical conductivity, making it suitable for applications in electronics and sensors.
(e) **Multifunctionality**: Inorganic nanocomposites often exhibit multifunctional properties, combining the recompenses of different materials. This makes them suitable for a wide range of applications, including energy storage, catalysis, sensing, biomedical devices, and environmental remediation.
(f) **Processing Techniques**: Various techniques are employed to fabricate inorganic nanocomposites, including solution casting, electrospinning, chemical vapor deposition, and 3D printing. The choice of technique depends on the desired composite structure, nanoparticle-matrix compatibility, and intended application.
(g) **Challenges**: The fabrication of inorganic nanocomposites involves challenges such as nanoparticle dispersion control, interface compatibility between nanoparticles and matrix, and maintaining the desired composite structure. Additionally, scaling up the production of nanocomposites while maintaining consistent properties can be a challenge.
(h) **Safety and Environmental Considerations**: As with any nanomaterial, it is important to consider the potential health and environmental impacts of inorganic nanocomposites. Proper handling, waste management, and risk assessment are essential to ensure their safe use and minimize any potential adverse effects.

Inorganic nanocomposites offer a wide range of opportunities for developing advanced materials with enhanced properties and tailored functionalities. Ongoing research and development in this field continue to explore new combinations of nanoparticles and matrix materials, as well as improved fabrication techniques, to unlock the full potential of inorganic nanocomposites for various applications.

4 Types of Inorganic Supported Nanocomposites

Here are a few examples of inorganic supported nanocomposites.

(a) **Catalyst Supported Nanocomposites**: Inorganic nanoparticles, such as metal catalysts (e.g., platinum, palladium), can be supported on high-surface-area materials like activated carbon, zeolites, or metal oxides. These supported catalysts are widely used in heterogeneous catalysis for various chemical reactions, such as hydrogenation, oxidation, and pollutant degradation [18].

(b) **Photocatalyst Supported Nanocomposites**: Semiconductor nanoparticles, such as titanium dioxide (TiO_2) or zinc oxide (ZnO), are commonly supported on materials like silica, alumina, or graphene oxide to enhance their photocatalytic activity [19]. These supported photocatalysts find applications in water purification, air pollution control, and solar energy conversion.

(c) **Sensor Supported Nanocomposites**: Inorganic nanoparticles, such as metal oxides (e.g., tin oxide, zinc oxide), can be supported on substrates like silicon, alumina, or quartz to create gas sensors or biosensors [20]. The supported nanocomposites exhibit improved sensitivity and selectivity, making them suitable for environmental monitoring, healthcare diagnostics, and industrial safety applications.

(d) **Membrane Supported Nanocomposites**: Inorganic nanoparticles, such as metal or metal oxide nanoparticles, can be incorporated into polymer membranes to enhance their separation properties. The nanoparticles can provide improved selectivity, fouling resistance, or antibacterial properties to the membrane [21]. These supported nanocomposite membranes are used in water treatment, gas separation, and fuel cell applications.

(e) **Energy Storage Supported Nanocomposites**: Inorganic nanoparticles, such as lithium metal oxides or carbon nanotubes, can be supported on conductive substrates like carbon or graphene to create electrode materials for batteries or supercapacitors. The supported nanocomposites offer enhanced energy storage capacity, cycling stability, and rate capability [22].

(f) **Nanofluid Supported Nanocomposites**: Inorganic nanoparticles can be dispersed in a liquid medium, such as water or oil, along with stabilizing agents to create nanofluids. These nanofluids can be used for various applications like heat transfer enhancement, lubrication, or as functional fluids with improved optical, thermal, or electrical properties [23].

These are just a few examples of the diverse range of inorganic supported nanocomposites that are being explored for various applications. The choice of the supporting material depends on the desired properties, compatibility with the nanoparticles, and the targeted application. The combination of inorganic nanoparticles with a suitable support material allows for synergistic effects and tailored functionalities, making these nanocomposites highly versatile and valuable for a wide range of technological advancements.

5 Inorganic Nanocomposites Synthesis Process

The synthesis of inorganic nanoparticle composites involves incorporating inorganic nanoparticles into a matrix material to create a composite structure. The specific synthesis methods depend on the nanoparticles and matrix materials used. Here are some common techniques for synthesizing inorganic nanoparticle composites:

(a) **In Situ Synthesis**: In this method, nanoparticles are synthesized directly within the matrix material. For example, inorganic precursors can be added to a polymer or sol–gel solution, followed by a chemical reaction or reduction process to form nanoparticles within the matrix [24]. This approach ensures a uniform dispersion of nanoparticles throughout the composite.
(b) **Solution Mixing**: In this technique, pre-synthesized nanoparticles are mixed with the matrix material in a suitable solvent. The nanoparticles can be surface-modified or functionalized to enhance their compatibility with the matrix. After mixing, the solvent is evaporated or removed to obtain a solid composite material [25].
(c) **Sol–Gel Process**: The sol–gel method involves the conversion of a precursor solution into a solid gel-like network. In the case of nanoparticle composites, the nanoparticles are dispersed in the precursor sol–gel solution, which is then polymerized or cross-linked to form a composite [26]. This technique is commonly used for synthesizing ceramic or glass-based nanoparticle composites.
(d) **Melt Mixing**: Melt mixing involves dispersing nanoparticles within a molten matrix material, typically a polymer or a metal. High shear forces, such as in extrusion or twin-screw mixing, are applied to ensure uniform dispersion of nanoparticles in the molten state. The mixture is then cooled and solidified to obtain the composite [27].
(e) **Electrospinning**: Electrospinning is a technique used to fabricate nanofibers with high surface area. In the case of nanoparticle composites, nanoparticles are dispersed within a polymer solution, and an electric field is applied to create a jet of the solution. The solvent evaporates during flight, and the resulting nanofibers contain uniformly distributed nanoparticles [27].
(f) **Chemical Vapor Deposition (CVD)**: CVD is a vapor-phase synthesis method commonly used for growing thin films or coatings. In the case of nanoparticle composites, precursor gases containing the desired elements are introduced, and a chemical reaction occurs on a substrate or pre-existing nanoparticles to form the composite structure [28].
(g) **Template-Assisted Methods**: Templates can be used as scaffolds to guide the growth or assembly of nanoparticles within a matrix [29]. For example, porous templates or sacrificial templates can be filled with nanoparticle suspensions, followed by removal or dissolution of the template to leave behind a composite structure.

It is important to note that the choice of synthesis method depends on factors such as the nature of the nanoparticles, the matrix material, the desired composite structure, and the intended application. Optimization of synthesis parameters, including temperature, reaction time, precursor concentrations, and mixing techniques, is necessary to achieve desired nanoparticle dispersion and composite properties.

6 Wastewater Treatment Process

In the search for effective water treatment technologies, research has been conducted on the development and use of advanced and innovative materials. Using cheap and efficient products can reduce water treatment costs and help more people access safe water.

Nanocomposites have shown promise in various types of water treatment for the removal of toxic contaminants. Different inorganic nanocomposites are listed with respect to their degradation or quenching efficiency percentage in Table 1.

Some common types of water treatment methods where nanocomposites have been utilized for toxicity removal include.

(a) **Adsorption**: Nanocomposites can be designed with high surface area and specific adsorption properties to remove toxic contaminants from water. Functionalized nanocomposites, such as metal–organic frameworks (MOFs), carbon nanotubes, or graphene oxide, can selectively adsorb and remove contaminants like heavy metals, organic pollutants, and pharmaceuticals from water.

(b) **Filtration**: Nanocomposite-based membranes or filters can be used to effectively remove toxic substances from water. Membranes incorporating nanoparticles, such as zeolites, clay minerals, or metal oxides, can exhibit enhanced adsorption and filtration capabilities, effectively trapping contaminants while allowing clean water to pass through.

(c) **Catalysis**: Nanocomposites can act as catalysts to facilitate the degradation or transformation of toxic contaminants in water. Supported nanoparticles, such as titanium dioxide (TiO_2), can be immobilized on a substrate to form a nanocomposite catalyst. When exposed to light or other energy sources, these catalysts can promote the degradation of organic pollutants through photocatalytic or advanced oxidation processes.

(d) **Disinfection**: Nanocomposites can be used for water disinfection by effectively eliminating pathogens and microorganisms. Silver nanoparticles, for example, have antimicrobial properties and can be incorporated into nanocomposites to enhance their disinfection capabilities, providing a viable alternative to traditional disinfection methods.

(e) **Ion Exchange**: Nanocomposites can be designed with ion exchange properties to remove toxic ions from water. Ion exchange resins or nanoparticles embedded in a matrix can selectively bind and remove specific ions, such as heavy metals or radioactive elements, from contaminated water.

Further sections discuss the various types of nanomaterials that have been employed as inorganic-supported nanocomposites to address the problems and challenges existing in water treatment processes.

Table 1 Comparison of inorganic nanocomposites based on their degradation or quenching efficiency (%)

S. No.	Catalysts	Method	Toxic contaminants	Maximum degradation/quenching percentage (%)	Experimental time (min)	References
1	g-C_3N_4	Photocatalytic	Congo red, Methyl orange	1.96, 1.37	60	[30]
2	g-C_3N_4TPyZnBTPy			10.23, 5.54		
3	g-C_3N_4TPy$(ZnBTPy)_3$			38.86, 24.57		
4	g-C_3N_4TPy$(ZnBTPy)_5$			17.96, 12.37		
5	Ch-PANI–CuO	Photocatalytic (dark condition, light irradiation)	Methyl orange	78.6 (dark), 95.1 (light irradiation)	60	[31]
6	$[Mg_2(APDA)_2(H_2O)_3]\cdot 5DMA\cdot 5H_2O$	Sensing	Fe (III), Nitrofurazon, Nitrofurantoin, 2,6-dichloro-4-nitroaniline	$\approx$ 100, 97.2, 97.6, 99	60	[32]
7	OPH^{6His}/UiO-66-NH_2	Sensing	Methyl parathion	95.9	60	[33]
8	MnO_2@ZIF-8	Adsorption	Arsenic	93	1500	[34]
9	MnO_2/PANI	Photocatalytic	Methylene blue	94	120	[35]
10	MZIF-67	Adsorption	Methyl orange	92.25	420	[36]
11	Fe@N–C /PMS	Photocatalytic	Orange II	100	120	[37]
12	Co@N–C /PMS		Rhodamine B, Methylene blue, Orange II, Methyl violet, Methyl orange	84.1, 96.0,100, 97.4,97.5	120	
13	Ni@N–C /PMS		Orange II	47.2	120	

6.1 Zinc-Based Nanocomposite

Yu et al. synthesized O–g-C_3N_4@$(ZnBTPy)_n$ nanocomposites [30], in which a graphitic carbon nitride (g-C_2N_4)-based nanocomposite was created by incorporating a zinc terpyridine (ZnTPy) coordination polymer (CP) for use in photocurrent generation and in catalyzing the breakdown of organic dyes.

- **Preparation**: The preparation process is divided into two steps, one is to modify-C_3N_4 nanoparticles by TPy coordination, and the other is to use Zn^{2+} ions as carriers to assemble ZnBTPy CP multilayer films on the surface of Zn^{2+} oriented LBL and O–g- C_3N_4 nanoparticles.

Wastewater Treatment

- **Photocatalytic Activity**: The catalytic performance was assessed using two organic dyes as model compounds: methyl orange (MO) and Congo red (CR). Both dyes exhibit intense absorption in the visible wavelength region, allowing spectroscopic methods to be employed for monitoring the degradation processes. Using those absorption spectra, the concentration of the dye was measured

Meaning of the relative C_t/C_0 concentration of CR in aqueous solution compared to irradiation time (t), where C_0 and C_t are the initial and CR concentration after irradiation. Using the first-order fitting equation, $\ln(Ct/C0) = kt$, the decay constant k was determined.

The increased decay rate can be attributed to the improved photocurrent generation; that is, essentially (1) both O–g-C_3N4 and ZnBTPy CPs act as photosensitizers and increase their absorption capacity; (2) in the binary nanocomposite, the charge transfer efficiency is reliable, which facilitates the separation and exchange of electron–hole pairs.

6.2 Copper Oxide Nanocomposite

Rathore et al. [31] synthesized a chitosan–polyaniline–copper(II) oxide (Ch–PANI–CuO) nanocomposite using the batch adsorption method. The photocatalytic performance of the composite was evaluated by studying the degradation of methyl orange dye under UV irradiation.

- **Preparation**: In this experiment, chitosan was dissolved in aqueous acetic acid, followed by the addition of aniline dissolved in acetic acid and ammonium peroxodisulfate. The resulting composite underwent washing and drying, resulting in a colorless material. Separately, copper chloride and copper oxide were dissolved in water and added to the chitosan-aniline composite, which was then washed and dried. The resulting composite was used as an adsorbent for dye removal.
- **Wastewater Treatment**

(1) **Adsorption Process**: The adsorption of methyl orange dye on the Ch–PANI–CuO nanocomposite was studied by analyzing different initial dye concentrations, pH levels, and adsorbent doses. The equilibrium time for adsorption was determined through the time course analysis, showing that the initial removal of the dye was rapid within the first 5 min and gradually decreased over time until reaching equilibrium. The rate of the removal of the methyl orange dye varied with various initial concentrations of the dye. The higher initial adsorption activity was attributed to the availability of more adsorption sites on the nanocomposite surface. The equilibrium condition for methyl orange was achieved after approximately 45 min of contact time. Additionally, the removal efficiency of methyl orange increased from 95.1% to 98.9% as the adsorbent dosage increased from 0.1 to 0.5 g L^{-1}, indicating the impact of concentration gradient between the dye and the adsorbent surface.

(2) **Photocatalytic Activity**: The photocatalytic activity of the Ch–PANI–CuO nanocomposite was evaluated to assess the impact of irradiation on the interaction between the catalyst and the dye. Two sets of experiments were conducted: one in the dark condition and the other with light irradiation. In the dark condition, the decolorization efficiency observed was 78.6%, while in the presence of light irradiation, the efficiency increased to 95.1%. This indicates that the nanocomposite exhibits enhanced photocatalytic activity when exposed to light, leading to higher dye removal efficiency.

6.3 Magnesium-Based Nanocomposite

Xu and colleagues designed and synthesized an amino-functionalized bridging ligand, H_2APDA, which reacts with Mg(II) ions to form the alkaline earth LMOF, noted as $[Mg_2(APDA)_2(H_2O)_3]\cdot 5DMA\cdot 5H_2O$ (Mg-APDA) [32]. The Mg-APDA structure consists of a three-dimensional framework featuring one-dimensional hexagonal channels. This structure uniquely functions as a selective fluorescent sensor for Fe(III) ions and effectively detects pesticides and antibiotics, a notable first for Mg-based LMOFs.

- **Preparation**: The synthesis of Mg-LMOF involves the reaction of an amino-decorated bridging ligand H_2APDA with Mg(II) ions. The ligand is first synthesized and then reacted with Mg(II) ions to construct the Mg-LMOF. The resulting Mg-LMOF is a three-dimensional framework along with one-dimensional hexagonal channels decked with Lewis base amino sites and uncoordinated O atoms, facilitating the binding and recognition of various analytes by Mg-APDA. Mg-LMOF can be used as a multi-response luminescence sensor for the efficient detection of Fe(III)ions, pesticides, and antibiotics.
- **Wastewater Treatment**

(1) **Sensing of Fe (III) Ions**: To study the presence of Fe(III) ions, fluorescence titration experiments were performed by incrementally adding a 1 mM DMF

solution of Fe(III) ions to a stable suspension of Mg-APDA in DMF. The mixture was then transferred into a quartz cell for analysis. Fluorescence measurements were conducted at room temperature, with emission wavelengths between 360 and 550 nm, and an excitation wavelength (λex) set at 351 nm. The Fe(III) ions displayed a UV − vis absorption spectrum that significantly overlapped with the emission spectrum of Mg-APDA, indicating an efficient energy transfer mechanism between the two. The significant quenching efficiency for Fe(III) ions is probably attributed to a combination of electron transfer induced by light and resonance energy transfer mechanisms.

(2) **Sensing of Pesticides**: Fluorescence titration experiments were conducted to detect various pesticides (nitrofurazone, nitrofurantoin, and 2,6-dichloro-4-nitroaniline) using a 1 mM DMF solution. The pesticides were added incrementally to a stable suspension of Mg-APDA in DMF, and the mixture was then transferred to a quartz cell for analysis. Fluorescence measurements were conducted at room temperature, using emission wavelengths from 360 to 550 nm and an excitation wavelength (λex) of 351 nm.

The sensitivity of Mg-APDA in detecting trace levels of pesticides was highlighted by the low detection limits observed (150 ppb for DCN, 108 ppb for NZF, and 126 ppb for NFT). This material demonstrated high sensitivity in identifying minute amounts of pesticides. To the authors' knowledge, this is the first reported instance of a fluorescence sensing material derived from a Mg-MOF that is capable of detecting low concentrations of pesticides effectively.

(3) **Sensing of Antibiotics**: Fluorescence titration experiments were conducted to detect different antibiotics (nitrofurazone and nitrofurantoin) using a 1 mM DMF solution. Gradual additions of the antibiotics were made to a stable suspension of Mg-APDA in DMF, and the resulting mixture was placed in a quartz cell for testing. Fluorescence detections were performed at room temperature, with emission wavelengths ranging from 360 to 550 nm and an excitation wavelength (λex) of 351 nm. The sensitivity of Mg-APDA in detecting trace levels of antibiotics was evident from the low detection limits observed (108 ppb for NZF and 126 ppb for NFT). This material demonstrated high sensitivity in identifying minute amounts of antibiotics. Importantly, this is the first reported instance of a Mg-MOF-based fluorescence sensing material capable of efficiently detecting trace quantities of antibiotics, according to the authors' knowledge.

6.4 Zirconium-Based Nanocomposite

J. Mehta et al. report the first instance of covalent immobilization of the OPH^{6His} enzyme onto UiO-66-NH_2, a type of zirconium metal–organic framework (MOF) [33]. The resulting bioconjugate, OPH^{6His}/UiO-66-NH_2, serves as a molecular carrier for the catalytic conversion of methyl parathion into p-nitrophenol, which can be detected by adding Coumarin1 as a fluorescent reporter. This work underscores the continued use of methyl parathion in developing countries such as India, despite the bans enacted by various government agencies.

- **Preparation**: The enzyme–MOF composite was created by covalently attaching the hexahistidine-tagged organophosphorus hydrolase (OPH^{6His}) to a zirconium metal–organic framework known as UiO-66-NH_2. This composite served as a carrier to enhance the hydrolysis of the organophosphate pesticide methyl parathion into p-nitrophenol (PNP). The production of PNP was directly proportional to the concentration of the pesticide added. To detect the presence of methyl parathion, Coumarin1, a fluorescent reporter molecule, was introduced into the reaction mixture. Since PNP had the effect of quenching the fluorescence of Coumarin, the detection of methyl parathion was possible over a broad concentration range of 10–10^6 ng/mL, with an achievable limit of quantification at 10 ng/mL.
- **Wastewater Treatment**

  Sensing of Methyl Parathion: The sensing of methyl parathion was carried out using an enzyme-MOF composite. The OPH^{6His} enzyme was immobilized on a Zr-MOF called UiO-66-NH_2 to facilitate the hydrolysis of methyl parathion into PNP. Coumarin1 was introduced as a reporter fluorescent molecule, and PNP was formed in direct proportion to the added pesticide concentration. The fluorescence of Coumarin1 was quenched by PNP, allowing for the detection of methyl parathion over a wide concentration range. The OPH^{6His}/UiO-66-NH_2 bioconjugate was found to be stable and reusable for at least eight times.

6.5 Manganese Dioxide Nanocomposite

6.5.1 Jian et al. developed a surfactant-free method to synthesize 1D MnO_2@MOFs nanowires by combining water-stable ZIF-8 nanoparticles with β-MnO_2 nanowires [34]. By controlling the dosage of MnO_2 nanowires and the solution medium, they successfully obtained well-defined MnO_2@ZIF-8 nanostructures. These unique composites showed enhanced adsorption capacity for As(III) removal compared to pristine ZIF-8 particles and demonstrated excellent settleability in water under gravity alone.

- **Preparation**: The β-MnO_2@ZIF-8 nanocomposite was prepared by a facile and rapid method in a methanol medium. The synthesis process involved the following steps:

  (1) Ultrathin tetragonal phased β-MnO_2 nanowires were prepared via a hydrothermal reaction.
  (2) A methanol solution of $Zn(NO_3)_2$was quickly added to the previously prepared mixture.

(3) The solution was stirred continuously for 2 h at room temperature to obtain well-defined MnO_2@ZIF-8 nanowires.
(4) The methanol present in the reaction solution was essential in reducing interfacial energy and aiding the integration process, leading to the formation of ZIF-8 nanocrystals on β-MnO_2 nanowires.
(5) By controlling the concentration of the ZIF-8 precursor, it was possible to adjust the thickness of the ZIF-8 shell surrounding the β-MnO_2 nanowires.

- **Wastewater Treatment**

Adsorption of Arsenic: The β-MnO_2@ZIF-8 nanocomposite was found to be effective in the adsorption of arsenic from water. The nanocomposite was able to convert As(III) to As(V), demonstrating enhanced adsorption for arsenite removal in comparison to bulk ZIF-8. The initial concentration of As(III) varied from 5 mg L^{-1} to 130 mg L^{-1}. The maximal adsorption capacity of As(III) was found to be 140.27 mg g^{-1} at room temperature. The kinetic and isotherm data for As(III) adsorption onto β-MnO_2@ZIF-8 nanocomposites aligned well with the pseudo-second-order and Langmuir models, respectively. The nanocomposite demonstrated a high settling rate under gravity following adsorption, highlighting its potential for practical applications.

6.5.2 The synthesis of a PANI-modified MnO_2 composite by H. Xu et al. demonstrates its remarkable catalytic activity in the degradation of dyes, namely methylene blue, methyl orange, and rhodamine B, under the gentle embrace of visible light irradiation [35].

- **Preparation**: MnO_2 composite nanorods were prepared through an interfacial polymerization method. The process included mixing 6 mmol of $KMnO_4$ and 9 mmol of $MnSO_4 \cdot H_2O$ in 40 mL of deionized water, followed by sonication for 30 min. This well-mixed solution was then added to 10 mL of carbon tetrachloride (CCl_4) inside a stainless-steel autoclave. The autoclave was sealed and maintained at 140 °C for 4 h. After cooling to room temperature, the precipitate was collected by filtration, rinsed with distilled water, and dried at 60 °C for 12 h.
- **Wastewater Treatment**

Photocatalytic Activity of Dyes: The photocatalytic performance of the polyaniline-modified MnO_2 composite nanorods was assessed through the degradation of organic dyes, specifically methylene blue, methyl orange, and rhodamine B, under visible light. To test the photocatalytic activity, visible light was used, and light below 420 nm was removed using a filter. The experimental setup involved mixing an aqueous solution of organic dyes with a concentration of 5 mg/L with 10 mg of catalyst and 10 mL of H_2O_2 (30%). The resulting suspension was stirred for 30 minutes in the dark to achieve adsorption-desorption equilibrium of the organic dyes. The change in the concentration in the solution of organic dyes was monitored at 15-minute intervals using a UV-vis spectrophotometer. The characteristic absorption peaks at 664 nm, 462 nm, and 554 nm for methylene blue, methyl orange, and rhodamine B solutions, respectively, were selected as indicators of the photocatalytic degradation process.

The findings demonstrated that the polyaniline-modified MnO_2 composites exhibited superior photocatalytic efficacy compared to that of polyaniline and MnO_2 individually. This excellent catalytic performance of the composite was attributed to synergistic effect of PANI and MnO_2 in the presence of H_2O_2.

6.6 *Iron-Based Nanocomposite*

6.6.1 In this study, Q. Yang et al. embarked upon the utilization of ZIF-67, an exceptionally captivating MOF material renowned for its diverse applications [36]. Their endeavor focused on the fabrication of an innovative class of porous magnetic MOF composites named Fe_3O_4-PSS@ZIF-67, which refers to as MZIF-67. Through comprehensive investigations, the adsorption properties of MZIF-67 toward methyl orange (MO) and methylene blue (MB) were explored, shedding light on their intriguing interactions.

- **Preparation**: To prepare the Fe_3O_4@MOFs magnetic porous composite material called MZIF-67, the authors used a solvothermal reduction method to prepare magnetic Fe_3O_4 particles. Then, they added 0.10 g of prepared Fe_3O_4 particles to 300 ml of an aqueous solution of PSS (30 wt%) under ultrasonic condition for 20 min. The samples were recovered by an external magnetic field and washed three times with distilled water. To grow ZIF-67 shell on the surface of Fe_3O_4 nanoparticles, an anionic polymer PSS was employed. Depending on the reaction conditions, the thickness of ZIF-67 would be adjusted. The resulting MZIF-67 composites are core–shell structures, with the aggregation core of Fe_3O_4 nanoparticles coated with petal-like ZIF-67 crystals.
- **Wastewater Treatment**

Adsorption of Dyes: The paper discusses the adsorption of methyl orange (MO) dye by the Fe_3O_4@MOFs magnetic porous composite material called MZIF-67. The MZIF-67 composites perform well on MO adsorption, with a maximum adsorption capacity of 738 $mg.g^{-1}$. The high adsorption capacity could be attributed to the highly porous nature and the Lewis base of coordinated Co^{2+} centrals. The pseudo-second-order constants were calculated to determine the adsorption rate constant. The paper also shows the selective separation of MO dye from the mixture dye solution containing MO and MB, with a removal rate of MO up to 92%.

6.2.2 In this study, Y. Yao et al. present a straightforward, cost-effective, and scalable synthesis method for the production of M@N–C through simple thermal treatment [37]. The obtained carbon materials undergo thorough physical and chemical characterizations; their catalytic efficiency is evaluated in the activation of oxidants peroxymonosulfate or peroxydisulfate (PMS or PDS) for the removal of organic dyes. Furthermore, a thorough analysis of the catalysts and experimental findings provides a detailed explanation of the mechanism involved in the M@N–C activation of PMS. The synthesized M@N–C particles exhibit enhanced catalytic performance and stability, which can be ascribed to the synergistic interaction between N-doped carbon nanotubes and M nanoparticles.

- **Preparation:** The paper reports the preparation of M-based nanocomposite material called M@N–C (M = Fe, Co, Ni) which consists of Fe nanoparticles encapsulated in nitrogen-doped carbon nanotubes. The synthesis was carried out using a simple one-pot hydrothermal method. The resulting product was then washed with deionized water and ethanol to remove any impurities and dried at 60 °C. The Fe@N–C nanocomposite was characterized using various techniques such as X-ray diffraction, transmission electron microscopy, and Raman spectroscopy. The results showed that the Fe nanoparticles were uniformly distributed within the nitrogen-doped carbon nanotubes, and the nanocomposite exhibited excellent catalytic activity for the degradation of organic pollutants.
- **Wastewater Management**

Photocatalytic Activity of Orange II: Fe nanocomposite's catalytic activity was assessed for organic pollutant removal. Its efficiency in activating oxidants for decolorization of organic dyes was tested, using the degradation of Orange II as a model reaction. Results indicated the Fe nanocomposite displayed remarkable catalytic activity, with approximately 100% of the dye degraded within 120 minutes in the activation of PMS. The involvement of hydroxyl radicals and sulfate radicals in the catalytic oxidation process was confirmed by utilizing different radical scavengers.

6.7 Cobalt-Based Nanocomposite

The Co@N-C nanocomposite was synthesized, and its catalytic efficiency was evaluated in activating oxidants, peroxymonosulfate or peroxydisulfate (PMS or PDS), to facilitate the removal of organic dyes [37].

- **Preparation:** 1. Preparation of Co-based metal–organic frameworks (MOFs) using cobalt nitrate and 2-aminoterephthalic acid as precursors.
- Carbonization of the Co-based MOFs to obtain Co-based nanocrystals encapsulated in carbon.
- Nitrogen doping of the carbon matrix by annealing the Co-based nanocrystals in the presence of ammonia gas to obtain Co-based nanocrystals encapsulated in nitrogen-doped carbon nanotubes.
- **Wastewater Management**

Photocatalytic Activity of Orange II: Co@N-C nanocomposite showed high efficiency for PMS activation, making it a promising catalyst for pollution control. Its catalytic activity with PDS and H_2O_2 was comparatively lower. The synergistic effect of M NPs and N-doped carbon contributed to its outstanding performance. With good catalytic activity, chemical stability, and easy recycling, Co@N-C nanocomposite holds potential for diverse environmental applications.

6.8 Nickel-Based Nanocomposite

Ni@N-C nanocomposite and their catalytic efficiency is assessed in the activation of oxidants (PMS, PDS, and H_2O_2) for the decolorization of organic dyes [37].

- **Preparation**: As the same process discussed above for the Co@N–C, Ni@N–C has been synthesized.
- **Wastewater Management**

  Photocatalytic Activity of Orange II: The paper reports that Ni@N-C nanocomposite was tested under the same conditions as Co@N-C and Fe@N-C catalysts for the activation of oxidants (PMS, PDS, and H_2O_2) for decolorization of organic dyes.

7 Conclusions

In conclusion, this chapter explored the fascinating realm of inorganic supported nanocomposites and their remarkable applications in removing toxic contaminants from wastewater. The synthesis and characterization of these nanocomposites were discussed, highlighting their unique properties and capabilities. The diverse range of toxic contaminants targeted for removal was examined, showcasing the versatility and effectiveness of inorganic supported nanocomposites in tackling various pollutants. The chapter also delved into the mechanisms involved in the removal process, shedding light on the underlying principles driving their exceptional performance. Overall, the findings underscore the immense potential of inorganic supported nanocomposites as a sustainable and efficient solution for addressing the challenges posed by toxic contaminants in wastewater, paving the way for cleaner and healthier environments.

Acknowledgements Dr. Kafeel Ahmad Siddiqui is thankful to the Council of Scientific and Industrial Research, New Delhi, India for awarding a research grant (01(2962)/18/EMR-II dated 01/05/2018) to support his "Crystal Engineering" Research.

References

1. Inyinbor Adejumoke A, Adebesin Babatunde O, Oluyori Abimbola P, Adelani Akande Tabitha A, Dada Adewumi O, Oreofe Toyin A (2018) Water pollution: effects, prevention, and climatic impact. Water Challenges Urbanizing World 33:33–47
2. Denchak M (2018) Water pollution: everything you need to know. Nat. Resour. Def. Counc, NY
3. Lin L, Yang H, Xu X (2022) Effects of water pollution on human health and disease heterogeneity: a review. Front Environ Sci 30(10):880246

4. Yang Y, Xu R, Ma CJ, Vlot AC, Klessig DF, Pichersky E (2008) Inactive methyl indole-3-acetic acid ester can be hydrolyzed and activated by several esterases belonging to the At MES esterase family of Arabidopsis. Plant Physiol 147(3):1034–1045
5. Harremoës P (1982) Immediate and delayed oxygen depletion in rivers. Water Res 16(7):1093–1098
6. Turner RE, Rabalais NN, Fry B, Atilla N, Milan CS, Lee JM, Normandeau C, Oswald TA, Swenson EM, Tomasko DA (2006) Paleo-indicators and water quality change in the Charlotte Harbor Estuary (Florida). Limnol Oceanography. 51(1part2):518–533
7. Obaideen K, Shehata N, Sayed ET, Abdelkareem MA, Mahmoud MS, Olabi AG (2022) The role of wastewater treatment in achieving sustainable development goals (SDGs) and sustainability guideline. Energy Nexus. 1(7):100112
8. Al-Tohamy R, Ali SS, Li F, Okasha KM, Mahmoud YA, Elsamahy T, Jiao H, Fu Y, Sun J (2022) A critical review on the treatment of dye-containing wastewater: Ecotoxicological and health concerns of textile dyes and possible remediation approaches for environmental safety. Ecotoxicol Environ Saf 1(231):113160
9. Prieto AL, Sigtermans LH, Mutlu BR, Aksan A, Arnold WA, Novak PJ (2016) Performance of a composite bioactive membrane for H 2 production and capture from high strength wastewater. Environ Sci: Water Res Technol 2(5):848–857
10. Back JO, Obholzer T, Winkler K, Jabornig S, Rupprich M (2018) Combining ultrafiltration and non-thermal plasma for low energy degradation of pharmaceuticals from conventionally treated wastewater. J Environ Chem Eng 6(6):7377–7385
11. Yaqoob AA, Parveen T, Umar K, Mohamad Ibrahim MN (2020) Role of nanomaterials in the treatment of wastewater: a review. Water 12(2):495
12. Saleh TA, Parthasarathy P, Irfan M (2019) Advanced functional polymer nanocomposites and their use in water ultra-purification. Trends Environ Analyt Chem 1(24):e00067
13. Debnath B, Biswas NT, Baidya R, Ghosh SK (2014) Nanotechnology in waste water treatment: a review. Ecol Urban Areas. 9:563
14. Kaur H, Devi N, Siwal SS, Alsanie WF, Thakur MK, Thakur VK (2023) Metal–organic framework-based materials for wastewater treatment: superior adsorbent materials for the removal of hazardous pollutants. ACS Omega 8(10):9004–9030
15. Rando G, Sfameni S, Galletta M, Drommi D, Cappello S, Plutino MR (2022) Functional Nanohybrids and Nanocomposites development for the removal of environmental pollutants and bioremediation. Molecules 27(15):4856
16. Gleick PH (1993) Water in crisis. Oxford University Press, New York
17. Wwap U. WWAP (United Nations World Water Assessment Programme)
18. Liu F, Diercks D, Hussain AM, Dale N, Furuya Y, Miura Y, Fukuyama Y, Duan C (2022) Nanocomposite catalyst for high-performance and durable intermediate-temperature methane-fueled metal-supported solid oxide fuel cells. ACS Appl Mater Interfaces 14(48):53840–53849
19. Suresh R, Gnanasekaran L, Rajendran S, Soto-Moscoso M, Chen WH, Show PL, Khoo KS (2023) Application of nanocomposites in integrated photocatalytic techniques for water pollution remediation. Environ Technol Innov 1(31):103149
20. Munonde TS, Nomngongo PN (2020) Nanocomposites for electrochemical sensors and their applications on the detection of trace metals in environmental water samples. Sensors 21(1):131
21. Corobea MC, Muhulet O, Miculescu F, Antoniac IV, Vuluga Z, Florea D, Vuluga DM, Butnaru M, Ivanov D, Voicu SI, Thakur VK (2016) Novel nanocomposite membranes from cellulose acetate and clay-silica nanowires. Polym Adv Technol 27(12):1586–1595
22. Chen S, Skordos A, Thakur VK (2020) Functional nanocomposites for energy storage: chemistry and new horizons. Mater Today Chem 1(17):100304
23. Nguele R, Nono KN, Sasaki K (2021) Nanocomposite and nanofluids: towards a sustainable carbon capture, utilization, and storage. In: Advances in Microfluidics and Nanofluids 2021 Jan 22. IntechOpen
24. Adnan MM, Dalod AR, Balci MH, Glaum J, Einarsrud MA (2018) In situ synthesis of hybrid inorganic–polymer nanocomposites. Polymers 10(10):1129

25. Zeng XF, Kong XR, Ge JL, Liu HT, Gao C, Shen ZG, Chen JF (2011) Effective solution mixing method to fabricate highly transparent and optical functional organic–inorganic nanocomposite film. Ind Eng Chem Res 50(6):3253–3258
26. Ramzan M, Kumar B, Usmani GA. Some Studies on Alumina-Zirconia Nanocomposite Obtained by Sol-gel Synthesis
27. Kango S, Kalia S, Thakur P, Kumari B, Pathania D (2015) Semiconductor–polymer hybrid materials. Organic-Inorganic Hybrid Nanomater 283–311
28. Hou X, Choy KL, Brun N, Serín V (2013) Nanocomposite coatings codeposited with nanoparticles using aerosol-assisted chemical vapour deposition. J Nanomater 2013(1):219039
29. Poolakkandy RR, Menamparambath MM (2020) Soft-template-assisted synthesis: a promising approach for the fabrication of transition metal oxides. Nanosc Adv 2(11):5015–5045
30. Yu HY, Li HJ, Ma YY, Feng YX, Qian DJ (2020) Interfacial self-assembly of carbon nitride-based nanocomposites with zinc terpyridyl coordination polymers for photocurrent generation and the photocatalytic degradation of organic dyes. Colloids Surf, A 5(596):124702
31. Rathore BS, Chauhan NP, Rawal MK, Ameta SC, Ameta R (2020) Chitosan–polyaniline–copper (II) oxide hybrid composite for the removal of methyl orange dye. Polym Bull 77(9):4833–4850
32. Xu N, Zhang Q, Hou B, Cheng Q, Zhang G (2018) A novel magnesium metal–organic framework as a multiresponsive luminescent sensor for Fe (III) ions, pesticides, and antibiotics with high selectivity and sensitivity. Inorg Chem 57(21):13330–13340
33. Jyotsana Mehta JM, Dhaka S, Paul AK, Siddavattam Dayananda SD, Akash Deep AD. Organophosphate hydrolase conjugated UiO-66-NH2 MOF based highly sensitive optical detection of methyl parathion
34. Jian Meipeng JM, Wang Huan WH, Liu RuiPing LR, Qu JiuHui QJ, Wang HuanTing WH, Zhang XiWang ZX. Self-assembled one-dimensional MnO_2 @ zeolitic imidazolate framework-8 nanostructures for highly efficient arsenite removal
35. Xu H, Zhang J, Chen Y, Lu H, Zhuang J, Li J (2014) Synthesis of polyaniline-modified MnO_2 composite nanorods and their photocatalytic application. Mater Lett 15(117):21–23
36. Yang Q, Ren S, Zhao Q, Lu R, Hang C, Chen Z, Zheng H (2018) Selective separation of methyl orange from water using magnetic ZIF-67 composites. Chem Eng J 1(333):49–57
37. Yao Y, Chen H, Lian C, Wei F, Zhang D, Wu G, Chen B, Wang S (2016) Fe Co, Ni nanocrystals encapsulated in nitrogen-doped carbon nanotubes as Fenton-like catalysts for organic pollutant removal. J Hazard Mater 15(314):129–139

Magnetic Nanocomposites and Their Applications in the Removal of Toxic Contaminants from Wastewater

Naseem Ahmad, Saimah Khan, Sanjana Yadav, Umme Salma, Nafish Fatima, Kashif Raees, Mohammad Shahadat, and Nafees Ahmad

Abstract The growing presence of water contaminants, such as organic pollutants and heavy metals, poses substantial environmental, and public health risks due to their toxicity and carcinogenic nature. Magnetic nanocomposites have emerged as effective materials for wastewater treatment, offering advantages like high surface area, magnetic responsiveness, and versatile functionalization capabilities. This chapter explores the synthesis techniques, such as co-precipitation, sol–gel, and green synthesis approaches, emphasizing their role in tailoring nanocomposites for enhanced performance and applications of magnetic nanocomposites in removing toxic contaminants from wastewater. Adsorption, photocatalytic degradation, and ion-exchange mechanisms are also highlighted in this chapter mentioning their effectiveness in removing toxic contaminants. Furthermore, it addresses the challenges associated with their practical application, including material stability, reusability, and potential environmental risks.

Keywords Magnetic nanocomposites · Heavy metals · Micropollutants · Adsorption · Advanced oxidation processes (AOPs)

N. Ahmad · S. Khan · S. Yadav · U. Salma · N. Fatima · N. Ahmad (✉)
Department of Chemistry, Integral University, Lucknow, India
e-mail: siddiquenafees123@gmail.com

K. Raees · M. Shahadat (✉)
Department of Chemistry, Chandigarh University, Mohali, Punjab, India
e-mail: mdshahadat93@gmail.com

N. Ahmad · S. Khan · S. Yadav · U. Salma · N. Fatima · K. Raees · M. Shahadat · N. Ahmad
Department of Textile and Fiber Engineering, Indian Institute of Technology, New Delhi, India

Institute of Oceanography and Environment, Universiti Malaysia Terengganu, Kuala Nerus, Terengganu, Malaysia

S. A. Bhawani and M. Jawaid (eds.), *Nanomaterials for Separation of Hazardous Contaminants from Wastewater*, Smart Nanomaterials Technology,
https://doi.org/10.1007/978-981-95-6079-0_7

1 Introduction

Water pollution is a pressing global issue, fueled by the rapid expansion of industries, urban areas, and populations, resulting in heightened pollutant release into water sources. Toxic pollutants like heavy metals, organic dyes, pharmaceuticals, pesticides, endocrine disruptors, and PFAS are increasingly found in wastewater, posing significant environmental concerns [1, 2]. These contaminants pose severe risks to aquatic ecosystems, human health, and biodiversity. Traditional wastewater treatment approaches, including coagulation, flocculation, sedimentation, and biological methods frequently fall short in effectively removing complex and persistent pollutants [3, 4]. The inadequacies of conventional wastewater treatment methods have spurred the development of advanced methods such as advanced oxidation processes (AOPs), membrane filtration, and adsorption, which offer improved contaminant removal efficiency [5]. Despite their effectiveness, advanced treatment methods are often hindered by high operational costs, energy demands, and potential secondary pollution. As a result, there is an urgent demand for cost-effective, efficient, and sustainable technologies to tackle the complex array of wastewater pollutants. Nanomaterials have transformed environmental remediation with their unique properties, including high surface area, adjustable pore sizes, and functionalization capabilities, enabling efficient adsorption and degradation of pollutants [6, 7]. These nanomaterials also possess distinct catalytic, magnetic, and optical properties, facilitating advanced applications like photocatalysis, Fenton-like reactions, and magnetic separation [8]. These properties make them versatile tools for tackling various classes of pollutants in wastewater. Compared to conventional materials, nanomaterials offer several advantages such as enhanced removal efficiency, removal of wide range of organic contaminants, functionalization enables the selective targeting of specific contaminants, reduced material usage and reusability. Many nanomaterials can be designed for multiple cycles of use, reducing costs and waste generation. These features position nanomaterials as cutting-edge tools in wastewater treatment, bridging the gap between efficiency, and sustainability [9].

Magnetic nanocomposites, a subclass of nanomaterials, have emerged as a promising solution in wastewater treatment, combining magnetic properties with multifunctional capabilities. These nanocomposites typically consist of magnetic cores, such as iron oxide nanoparticles, and functionalized shells or matrices, offering significant benefits [10]. Their magnetic properties enable easy recovery and reuse through external magnetic fields, minimizing material loss and secondary pollution. Magnetic nanocomposites can effectively remove diverse contaminants, including heavy metals and organic pollutants, via advanced oxidation, catalytic degradation, and adsorption. Their versatility also enables integration with other treatment methods, such as membrane filtration or photocatalysis, to create hybrid systems that boost overall efficiency [11]. Beyond removing pollutants, magnetic nanocomposites support sustainable wastewater treatment through their recyclability and reusability, promoting environmentally friendly practices. Furthermore, their adaptability to treat complex wastewater matrices makes them a promising candidate for addressing the challenges of modern wastewater treatment. Magnetic nanocomposites represent a significant step forward in developing advanced materials for environmental remediation.

2 Fundamentals and Properties of Magnetic Nanocomposites

Nanoparticle's small size and large surface area allow them to interact effectively with pollutants in wastewater, making them highly promising for water treatment applications. Their unique properties allow for efficient dispersion and strong interactions with contaminants, facilitating effective removal and degradation [12]. Despite being effective in removing pollutants from wastewater, nanoparticles can also acts as secondary pollutant. Nanoparticle accumulation and degradation can yield toxic materials, threatening ecosystems and human health. Therefore, it is essential to retrieve these nanoparticles after the removal of pollutants to prevent potential environmental and health risks. Researchers are working to develop efficient methods for nanoparticle removal from treated water, ensuring the water is safe for consumption or release back into the environment. To address these challenges, magnetic nanocomposites have emerged as a promising solution for environmental remediation. Their magnetic properties enable simple and effective retrieval from water bodies via magnetic separation, facilitating recovery and reuse [13]. Magnetic nanocomposites are highly effective in wastewater treatment due to their strong magnetic response, allowing for rapid and efficient recovery from suspensions. This property reduces environmental contamination risks and enables nanomaterial reuse, making them a valuable tool for sustainable water treatment. Another critical factor is their high contaminant removal rate, which is essential for effective wastewater remediation. Magnetic nanocomposites exhibit rapid and efficient adsorption of contaminants onto their surfaces, significantly reducing pollutant concentrations in water. Additionally, they facilitate the catalytic degradation of contaminants, breaking them down into harmless byproducts. This effective removal of pollutants improves water quality, making it safer for human consumption, industrial use, and environmental discharge. High removal rates also support the development of sustainable water remediation technologies, reducing environmental impact and promoting eco-friendly solutions. Another advantage is the ease of reuse of magnetic nanocomposites, after recovery, their catalytic properties remain intact, allowing for immediate reuse, which minimizes waste generation and conserves resources [14]. Furthermore, these materials demonstrate exceptional chemical stability, enabling them to withstand repeated use and environmental exposure without degradation. This stability is crucial, as unstable nanomaterials can release toxic byproducts, posing environmental and health risks [15]. Magnetic nanocomposites offer enhanced permeability for efficient pollutant removal and can be tailored for targeted removal of specific contaminants like heavy metals, pesticides, or industrial pollutants [16]. For example, studies show magnetic chitosan nanocomposites successfully extract heavy metals like cadmium, copper, iron, and nickel, demonstrating their potential in water purification [17]. Lastly, the extensive surface area of magnetic nanocomposites provides abundant binding sites for pollutants, boosting adsorption capacity, and facilitating chemical reactions that degrade contaminants, thereby enhancing wastewater treatment effectiveness [18]. The synergy of their properties makes magnetic nanoparticles and nanocomposites an attractive option for efficient and eco-friendly wastewater remediation.

These nanocomposites incorporate a diverse range of metal oxides, such as iron oxide, titanium oxide, zinc oxide, and others like manganese, aluminum, magnesium, cerium, and zirconium oxides [19]. In recent years, iron oxide (Fe_2O_3) nanomaterials have shown promise as an effective solution for eliminating toxic contaminants from wastewater [20]. Iron's abundance and the ease of synthesis iron oxides have sparked significant research into these materials. Key iron oxides under extensive study are hematite (α-Fe_2O_3), goethite (α-FeOOH), magnetite (Fe_3O_4), maghemite (γ-Fe_2O_3), and hydrous iron oxides (HFO) [21]. Notably, iron in goethite (α-FeOOH), hematite (α-Fe_2O_3), maghemite (γ-Fe_2O_3) exhibits a trivalent state, whereas in Fe_3O_4, iron displays both divalent and trivalent states. Goethite (α-FeOOH), a naturally occurring mineral, is a competitive and efficient adsorbent for heavy metals, boasting high adsorption capacity and environmental sustainability at a lower cost [22]. Hematite (α-Fe_2O_3) has demonstrated versatile application potential toward environmental monitoring, and catalysis. Additionally, hematite and maghemite (γ-Fe_2O_3) nanoparticles also show effectiveness in adsorbing heavy metals and other pollutants [23]. With their high surface area, maghemite and hematite nanoparticles exhibit exceptional adsorption capacity. Furthermore, their magnetic properties enable easy separation from treated wastewater using a magnetic field. Additionally, maghemite nanoparticles are also easy to synthesize, eco-friendly, and do not generate secondary pollution [24]. Magnetite (Fe_3O_4)-based nanomaterials are favored for adsorption applications due to their cost effectiveness, ease of use, and environmentally friendly properties [25]. Like maghemite, magnetite-based nanomaterials can be easily separated from water using a magnetic field, making them highly effective for removing heavy metals, as shown in numerous studies. Another class of magnetic nanoparticle, hydrous iron oxide (HFO) shows promise in removing contaminants like heavy metals, dyes, and pharmaceuticals from wastewater through adsorption, ion exchange, and co-precipitation. However, their poor mechanical stability and low hydraulic conductivity limit their use. To address this, HFO nanoparticles are often combined with porous materials to enhance their effectiveness [26]

3 Synthesis of Magnetic Nanocomposites

Magnetic nanocomposites are created by combining magnetic nanoparticles with a matrix, resulting in enhanced properties like magnetic behavior, thermal stability, and biocompatibility. Magnetic nanocomposites consist of two primary components: magnetic nanoparticles such as iron oxides (e.g., Fe_3O_4, γ-Fe_2O_3), ferrites

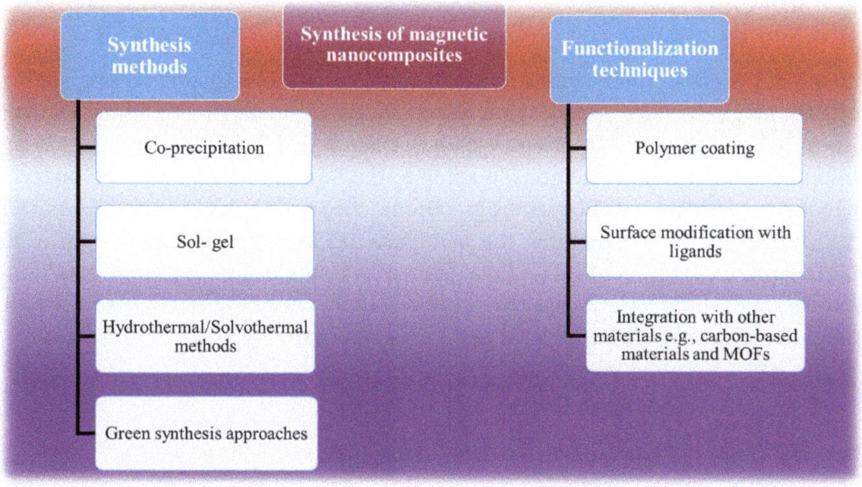

Fig. 1 Block diagram for the various methods for the synthesis of magnetic nanocomposites

(e.g., $CoFe_2O_4$, $MnFe_2O_4$), or metals (e.g., Fe, Co, Ni) and the matrix material to for the composites are polymers, silica, carbon-based materials (graphene, carbon nanotubes). Magnetic nanocomposites can be synthesized through several methods, including co-precipitation, sol–gel, hydrothermal/solvothermal, and green synthesis, each tailored to specific properties and applications. A block diagram is presented below in Fig. 1 for the various methods used for the synthesis of magnetic nanocomposites.

3.1 Co-precipitation

In the co-precipitation method, magnetic nanoparticles like Fe_3O_4 are created by simultaneously precipitating Fe^{2+} and Fe^{3+} ions in an alkaline solution (e.g., NaOH or NH_4OH). The magnetic nanocomposite is then formed by dispersing these nanoparticles in a matrix material, such as a polymer or silica. Kalantari et al. [27] synthesized talc/Fe_3O_4 nanocomposites at room temperature using ferric chloride ($FeCl_3$), ferrous chloride ($FeCl_2$), and sodium hydroxide (NaOH) as precursors. Co-precipitation yields ultra-fine iron oxide nanoparticles with reduced agglomeration, facilitating easy separation and paving the way for the development of novel magnetic nanocomposite materials.

3.2 Sol–Gel

The sol–gel process starts with preparing a sol by dissolving precursors, such as $FeCl_3$ for magnetic nanoparticles and alkoxides for the matrix material, in a solvent to form a homogeneous solution. Water is then added to initiate hydrolysis and subsequent condensation reactions, leading to the formation of the desired material. Magnetic nanoparticles can be incorporated into the sol through two methods: in situ or ex situ. In the in situ method, the nanoparticles are formed directly within the sol, allowing for their creation and integration in a single step. By adding reducing agents or controlling the pH and temperature, such as forming Fe_3O_4 nanoparticles through the co-precipitation of Fe^{2+} and Fe^{3+} ions [28].

3.3 Hydrothermal/Solvothermal Methods

The hydrothermal/solvothermal method involves reacting iron salt solutions or mixed metal hydroxides in aqueous media, leading to hydrolysis, oxidation, and ultimately the formation of ferrite materials. The solvothermal process takes place in a sealed autoclave reactor under high temperatures and autogenous pressure. This controlled environment boosts the solvent's capacity to dissolve precursors, promoting efficient reactions and accelerating the formation of the desired product. Several researchers have employed the solvothermal method to produce a range of magnetic materials such as $ZnFe_2O_4$, $NiFe_2O_4$, and $CuFe_2O_4$ etc. [29–31]. Magnetite nanoparticles are utilized for two key applications: photocatalytically breaking down organic pollutants and adsorbing heavy metals [32, 33]. The solvothermal method offers advantages like producing materials with high crystallinity, uniform particle size, and cost effective.

3.4 Green Synthesis Approaches

The green synthesis of Fe_3O_4 (iron oxide) nanoparticles and their composites is an eco-friendly approach that leverages natural materials to produce these nanoparticles. This method has been applied in various fields, including water treatment, and other areas and offers several benefits such as environmentally friendly that reduces reliance on hazardous chemicals, sustainability as it promotes the use of natural and biodegradable materials, cost effective method that utilizes plant extracts or other natural materials as reducing agents. In the green synthesis approach, the plant extracts or biodegradable materials are used as reducing agents. Several studies have been reported so far such as investigated the structure property relationships and chemistry of pollutant-magnetic material interactions [34–36].

4 Mechanisms of Toxic Contaminant Removal

The removal of toxic contaminants from water and other environmental matrices is a critical challenge in environmental science and engineering. Various mechanisms have been developed and employed to effectively remove or degrade these pollutants. Below, we discuss four key mechanisms: adsorption, catalytic degradation, ion exchange, and chelation, and the synergistic effects of magnetic nanocomposites.

4.1 *Adsorption*

Adsorption is a surface-based process where contaminants are removed from a solution by adhering to the surface of an adsorbent material. This adsorption mechanism is widely used due to its simplicity, efficiency, and cost effectiveness. Physical adsorption involves weak van der Waals forces between the adsorbate (contaminant) and the adsorbent (nanomaterials) surface. Chemical adsorption involves stronger chemical bonds, such as covalent or ionic interactions, between the adsorbate and adsorbent [37, 38]. In the electrostatic interactions, the charged contaminants are attracted to oppositely charged sites on the adsorbent surface. There studies can be described using the Langmuir isotherm model assumes monolayer adsorption on a homogeneous surface, while the Freundlich isotherm model describes multilayer adsorption on heterogeneous surfaces. These two models help understand the adsorption behavior and interactions between adsorbates and adsorbents. Several adsorbents such as nanomaterials, polymer nanocomposites, bionanocomposites, activated carbon, biochar, carbon rich sources, i.e., graphene oxide, MWCNTs, SWCNTs are widely used for contaminant removal [39]. A pictorial diagram is presented in Fig. 2 to show the adsorption of toxic contaminants by magnetic nanoparticles.

4.2 *Advanced Oxidation Processes (AOPs)*

AOPs involve advanced oxidation processes such as photocatalysis, ozonation, and electrochemical oxidation. The photo-Fenton process generates reactive oxygen species (ROS), like hydroxyl radicals. $^{\circ}OH$, to oxidize and degrade contaminants and superoxide radicals $^{\circ}O_2^{-}$ to reduce the toxic contaminants. The photocatalysis method utilizes semiconductors (e.g., such as nanomaterials, polymer nanocomposites, bionanocomposites, graphene oxide, MWCNTs, SWCNTs) under light irradiation to generate ROS [40]. Further, these ROS will react to the toxic contaminates and leads to the mineralization. The ozonation process used to directly oxidize contaminants or generate $\cdot OH$ radicals using O_3 [41]. In case of electrochemical oxidation which utilizes electrodes to generate ROS and degrade toxic pollutants, the photo-Fenton process combines Fenton's reagents, specifically iron ions (Fe^{2+})

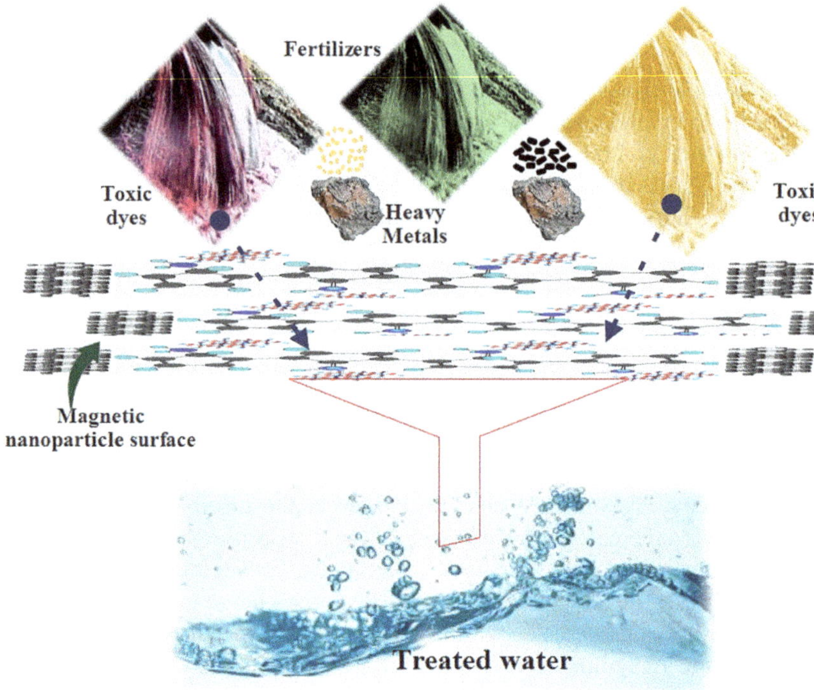

Fig. 2 Adsorption of various toxic contaminants on the surface of magnetic nanoparticle

and hydrogen peroxide (H_2O_2).with UV or visible light to enhance These hydroxyl radicals (·OH) are highly reactive and effectively degrade organic pollutants effectively degrade toxic organic contaminants in water. The light exposure regenerates Fe^{2+} from Fe^{3+}, sustaining the catalytic cycle and improving efficiency [42, 43]. This process requires optimized pH and controlled iron concentrations to prevent sludge formation. The photo-Fenton process is particularly effective for persistent pollutants like pharmaceuticals, pesticides, and dyes. Figure 3 shows the photocatalytic mechanism for breaking down harmful pollutants.

4.3 *Ion Exchange and Chelation Mechanisms*

These mechanisms are especially effective for removing heavy metals and other ionic contaminants from water. Ion exchange is a process where ions are reversibly swapped between a solid resin and a solution. Cations (e.g., Pb^{2+}, Cd^{2+}) or anions (e.g., $CrO_4{}^{2-}$) in the solution are replaced by harmless ions (e.g., Na^+, Cl^-) from the resin [38]. Commonly used resins include zeolites, clay minerals, and synthetic polymers. In Chelation, that involves forming stable, water-soluble complexes between

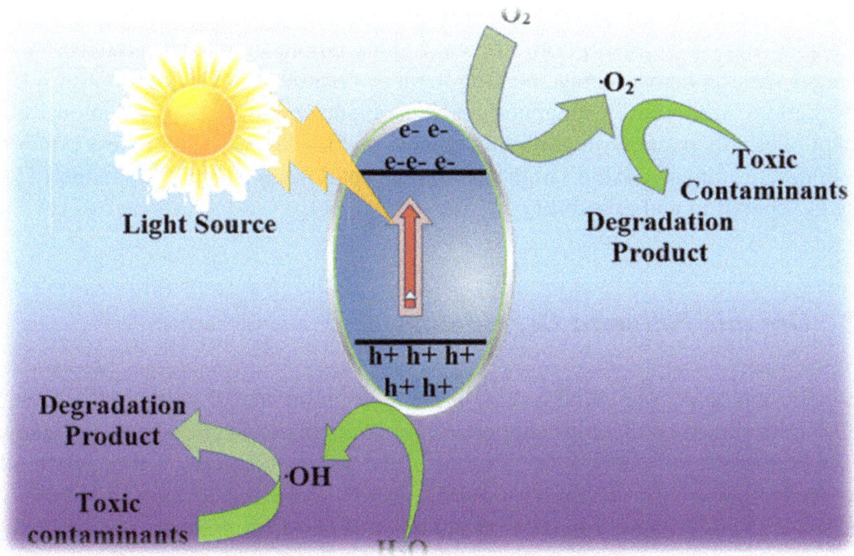

Fig. 3 Photocatalytic mechanism for the degradation of toxic contaminants

metal ions and chelating agents like EDTA or citric acid. These complexes can then be removed through precipitation or filtration. Chelation is highly effective for removing heavy metals from contaminated solution selective and effective for heavy metal removal [44].

5 Application of Magnetic Nanoparticle for the Removal of Toxic Contaminants

Magnetic nanoparticles have gained attention for environmental remediation, especially in water treatment, due to their unique properties like high surface area and ease of functionalization, making them effective for removing toxic contaminants., make them highly effective for the adsorption, separation, and degradation of pollutants.

5.1 Heavy Metal Removal

Magnetic nanoparticles (MNPs) particularly iron oxide nanoparticles like Fe_3O_4 have shown remarkable effectiveness in adsorbing heavy metals such as Pb, Cd, Hg, As, and Cr. The adsorption mechanisms involve surface complexation, electrostatic interactions, and ion exchange [45, 46]. Functionalizing MNPs with ligands such as thiols, amines, or carboxyl groups significantly enhances their affinity for heavy metals, leading to improved adsorption efficiency. This tailored approach enables MNPs to target specific heavy metals, making them a versatile solution for environmental remediation. Additionally, magnetic graphene oxide composites have demonstrated exceptional effectiveness in removing arsenic and lead from water. These composites integrate graphene oxide's high adsorption capacity with magnetic nanoparticles' ease of separation and reusability.

5.2 *Organic Pollutant Degradation*

Magnetic nanoparticles (MNPs) play a key role in advanced oxidation processes (AOPs), serving as catalysts or supports to break down persistent organic pollutants. These pollutants, including dyes, pesticides, and pharmaceuticals, can be efficiently degraded into less harmful compounds using MNPs. For instance, the combination of iron oxide (Fe_3O_4) nanoparticles with Fenton or photo-Fenton processes effectively breaks down organic contaminants into less toxic compounds. Similarly, magnetic nanoparticles (MNPs) coated with titanium dioxide (TiO_2) or other photocatalysts enable the photocatalytic degradation of pollutants when exposed to light. There are several benefits of using MNPs in AOPs such as high efficiency, easy magnetic recovery, and reusability [47, 48].

5.3 *Removal of Radioactive Contaminants*

Magnetic nanoparticles (MNPs) have been shown to efficiently adsorb radionuclides, including uranium, cesium, and strontium, through surface complexation and ion-exchange mechanisms such as in uranium removal MNPs coated with phosphates, silicates, or organic ligands demonstrate high efficiency in uranium removal. In the same way, Fe_3O_4-based composites are widely used to adsorb cesium and strontium from nuclear wastewater. MNPs offer high selectivity for target radionuclides, stability under harsh conditions, and with the added benefit of easy separation using an external magnet. Overall, MNPs provide a reliable and efficient solution for the remediation of radioactive contaminants in environmental and nuclear waste management applications [49, 50].

5.4 *Pathogen and Microbial Contaminant Removal*

Magnetic nanoparticles (MNPs) have been observed to effectively adsorb or inactivate bacteria, viruses, and other pathogens, making them a valuable asset for purifying water [51, 52]. Research has demonstrated that silver-coated MNPs exhibit strong antimicrobial properties, enabling removing pathogens from water. Additionally, functionalized MNPs can effectively eliminate specific microorganisms. There are several benefits of using MNPs in water purification such as rapid and efficient pathogen removal, potential for disinfection, and easy magnetic separation. MNPs hold great promise in addressing microbial contamination in water treatment, ensuring safe drinking water. Their unique properties make them an ideal solution for water purification applications. A table presented below summarizing different methods for removing toxic pollutants from wastewater, including their advantages and disadvantages (Table 1).

6 Conclusion and Future Prospects

Magnetic nanocomposites have demonstrated significant potential for the wastewater treatment which offers innovative solutions to combat water pollution, thanks to their unique blend of magnetic properties, photocatalytic activity, high adsorption capacity, and large surface area. This combination enables efficient removal of a wide range of carcinogenic contaminants, including pharmaceuticals, personal care products (PPCPs), heavy metals, fertilizers, and dyes, making them a promising tool for water purification. The magnetic nanoparticles are key to advancing wastewater treatment technologies through enhancing efficiency, reducing material cost and enabling easy recovery and recyclability of the material. Sustainable approaches to the design and use of magnetic nanoparticles are critical for the long-term application in wastewater treatment. Therefore, advancement in material design and functionalization is needed to optimize the performance enabling higher selectivity and stability in complex wastewater system. The combination of magnetic nanocomposites with other material such as carbon rich material (graphene oxide, MWCNTs, SWCNTs) creates synergy that enhances overall treatment efficiency. Further, there is a growing need to develop eco-friendly synthesis methods for magnetic nanocomposites, aligning with green chemistry principles to minimize environmental impact. This approach ensures sustainable production while reducing harm to the ecosystem. Additionally, targeted research on the application of magnetic nanocomposites for the specific contaminants such as microplastics, nanoplastics, PFAS can address emerging challenges in wastewater treatment. Magnetic nanocomposites have the potential to revolutionize wastewater treatment methods and help create a cleaner, safer, and more sustainable future by bridging the gap between material science and environmental sustainability.

Table 1 Summarizing various methods for the removal of toxic contaminants in wastewater with advantage and disadvantages

Contaminant	Removal method	Advantages	Disadvantages
Heavy metals	Adsorption (e.g., MNPs)	High efficiency, easy separation, reusable, cost effective for small-scale use	Potential nanoparticle toxicity, limited scalability, requires functionalization
	Ion exchange	High selectivity, effective for low concentrations, regenerable resins	High operational cost, sensitive to pH, resin fouling
	Chemical precipitation	Simple, cost effective, suitable for high concentrations	Sludge generation, requires pH adjustment, ineffective for low concentrations
Organic pollutants	Advanced oxidation processes (AOPs)	Degrades persistent pollutants, no sludge generation, versatile	High energy consumption, costly chemicals, potential byproduct toxicity
	Activated carbon adsorption	High adsorption capacity, effective for a wide range of organics	Expensive, regeneration required, limited by pore clogging
	Biodegradation	Eco-friendly, cost effective, uses natural processes	Slow process, sensitive to environmental conditions, incomplete degradation
Radioactive contaminants	Magnetic nanoparticles (MNPs)	High selectivity, easy separation, stable under harsh conditions	High synthesis cost, potential environmental risks, limited large-scale use
	Co-precipitation	Simple, effective for uranium and radium removal	Generates radioactive sludge, requires disposal, limited to specific radionuclides
	Ion-exchange resins	High efficiency, regenerable, selective for specific radionuclides	Expensive, sensitive to pH, resin fouling
Pathogen and microbial contaminant removal	Magnetic nanoparticles (MNPs)	Rapid removal, reusable, potential for disinfection	High cost, potential toxicity, limited scalability

(continued)

Table 1 (continued)

Contaminant	Removal method	Advantages	Disadvantages
	Chlorination	Effective disinfection, low cost, widely used	Forms harmful disinfection byproducts (DBPs), ineffective against some pathogens
	UV irradiation	No chemical addition, effective against a wide range of pathogens	High energy consumption, no residual disinfection, limited by water turbidity

References

1. Li X, Shen X, Jiang W, Xi Y, Li S (2024) Comprehensive review of emerging contaminants: detection technologies, environmental impact, and management strategies. Ecotoxicol Environ Saf 15(278):116420
2. Kumar V, Sharma N, Umesh M, Chakraborty P, Kaur K, Duhan L, Sarojini S, Thazeem B, Pasrija R, Vangnai AS, Maitra SS (2023) Micropollutants characteristics, fate, and sustainable removal technologies for landfill leachate: a technical perspective. J Water Process Eng. 1(53):103649
3. Teh CY, Budiman PM, Shak KP, Wu TY (2016) Recent advancement of coagulation–flocculation and its application in wastewater treatment. Ind Eng Chem Res 55(16):4363–4389
4. Ahmed MB, Zhou JL, Ngo HH, Guo W, Thomaidis NS, Xu J (2017) Progress in the biological and chemical treatment technologies for emerging contaminant removal from wastewater: a critical review. J Hazard Mater 5(323):274–298
5. Rashid R, Shafiq I, Akhter P, Iqbal MJ, Hussain M (2021) A state-of-the-art review on wastewater treatment techniques: the effectiveness of adsorption method. Environ Sci Pollut Res 28:9050–9066
6. Jawed A, Saxena V, Pandey LM (2020) Engineered nanomaterials and their surface functionalization for the removal of heavy metals: a review. J Water Process Eng 1(33):101009
7. Li H, Chen X, Shen D, Wu F, Pleixats R, Pan J (2021) Functionalized silica nanoparticles: classification, synthetic approaches and recent advances in adsorption applications. Nanoscale 13(38):15998–16016
8. Thomas N, Dionysiou DD, Pillai SC (2021) Heterogeneous Fenton catalysts: a review of recent advances. J Hazard Mater 15(404):124082
9. Ahmad N, Rahman QI, Iqbal A, Siddiqui MR, Wabaidur SM, Ahmad N, Coulon F (2024) Design of novel Sb_2S_3/polythiophene heterojunction for efficient adsorption and methanol sensing. Sep Purif Technol 18(350):127826
10. Ahmad I, Aftab MA, Fatima A, Mekkey SD, Melhi S, Ikram S (2024) A comprehensive review on the advancement of transition metals incorporated on functional magnetic nanocomposites for the catalytic reduction and photocatalytic degradation of organic pollutants. Coord Chem Rev 1(514):215904
11. Zhang J, Wu H, Shi L, Wu Z, Zhang S, Wang S, Sun H (2024) Photocatalysis coupling with membrane technology for sustainable and continuous purification of wastewater. Sep Purif Technol 15(329):125225
12. Liu Y, Tourbin M, Lachaize S, Guiraud P (2014) Nanoparticles in wastewaters: hazards, fate and remediation. Powder Technol 1(255):149–156
13. Bushra R, Ahmad M, Alam K, Seidi F, Shakeel S, Song J, Jin Y, Xiao H (2024) Recent advances in magnetic nanoparticles: key applications, environmental insights, and future strategies. Sustain Mater Technol 20:e00985

14. Mehmood A, Khan FS, Mubarak NM, Tan YH, Karri RR, Khalid M, Walvekar R, Abdullah EC, Nizamuddin S, Mazari SA (2021) Magnetic nanocomposites for sustainable water purification—a comprehensive review. Environ Sci Pollut Res 28:19563–19588
15. Vallinayagam S, Rajendran K, Lakkaboyana SK, Soontarapa K, Remya RR, Sharma VK, Kumar V, Venkateswarlu K, Koduru JR (2021) Recent developments in magnetic nanoparticles and nano-composites for wastewater treatment. J Environ Chem Eng 9(6):106553
16. Saleh TA (2024) Materials, nanomaterials, nanocomposites, and methods used for the treatment and removal of hazardous pollutants from wastewater: treatment technologies for water recycling and sustainability. Nano Struct Nano Objects 1(39):101231
17. Namdeo M, Bajpai SK (2008) Chitosan–magnetite nanocomposites (CMNs) as magnetic carrier particles for removal of Fe (III) from aqueous solutions. Colloids Surf A 320(1–3):161–168
18. Yuwei C, Jianlong W (2011) Preparation and characterization of magnetic chitosan nanoparticles and its application for Cu (II) removal. Chem Eng J 168(1):286–292
19. Yang J, Hou B, Wang J, Tian B, Bi J, Wang N, Li X, Huang X (2019) Nanomaterials for the removal of heavy metals from wastewater. Nanomaterials 9(3):424
20. Saharan P, Chaudhary GR, Mehta SK, Umar A (2014) Removal of water contaminants by iron oxide nanomaterials. J Nanosci Nanotechnol 14(1):627–643
21. Hua M, Zhang S, Pan B, Zhang W, Lv L, Zhang Q (2012) Heavy metal removal from water/wastewater by nanosized metal oxides: a review. J Hazard Mater 15(211):317–331
22. Massalimov IA, Il'Yasova RR, Musavirova LR, Samsonov MR, Mustafin AG (2014) Use of micrometer hematite particles and nanodispersed goethite as sorbent for heavy metals. Russian J Appl Chem 87:1456–1463
23. Ahmed MA, Ali SM, El-Dek SI, Galal A (2013) Magnetite–hematite nanoparticles prepared by green methods for heavy metal ions removal from water. Mater Sci Eng B 178(10):744–751
24. Tuutijärvi T, Lu J, Sillanpää M, Chen G (2009) As (V) adsorption on maghemite nanoparticles. J Hazard Mater 166(2–3):1415–1420
25. Hao YM, Man C, Hu ZB (2010) Effective removal of Cu (II) ions from aqueous solution by amino-functionalized magnetic nanoparticles. J Hazard Mater 184(1–3):392–399
26. Iqbal Z, Tanweer MS, Alam M (2022) Recent advances in adsorptive removal of wastewater pollutants by chemically modified metal oxides: a review. J Water Process Eng 1(46):102641
27. Kalantari K, Ahmad MB, Shameli K, Khandanlou R (2013) Synthesis of talc/Fe_3O_4 magnetic nanocomposites using chemical co-precipitation method. Int J Nanomed 15:1817–1823
28. Jesus J, Regadas J, Costa B, Carvalho J, Pádua A, Henriques C, Soares PI, Gavinho S, Valente MA, Graça MP, Soreto TS (2024) Green sol-gel synthesis of iron oxide nanoparticles for magnetic hyperthermia applications. Pharmaceutics 16(12):1578
29. Fotukian SM, Barati A, Soleymani M, Alizadeh AM (2020) Solvothermal synthesis of $CuFe_2O_4$ and Fe_3O_4 nanoparticles with high heating efficiency for magnetic hyperthermia application. J Alloy Compd 5(816):152548
30. Sun H, Cao L, Lu L (2011) Magnetite/reduced graphene oxide nanocomposites: one step solvothermal synthesis and use as a novel platform for removal of dye pollutants. Nano Res 4:550–562
31. Yáñez-Vilar S, Sánchez-Andújar M, Gómez-Aguirre C, Mira J, Señarís-Rodríguez MA, Castro-García S (2009) A simple solvothermal synthesis of MFe_2O_4 (M= Mn, Co and Ni) nanoparticles. J Solid State Chem 182(10):2685–2690
32. Taghavi R, Rostamnia S, Farajzadeh M, Karimi-Maleh H, Wang J, Kim D, Jang HW, Luque R, Varma RS, Shokouhimehr M (2022) Magnetite metal–organic frameworks: applications in environmental remediation of heavy metals, organic contaminants, and other pollutants. Inorg Chem 61(40):15747–15783
33. Habila MA, ALOthman ZA, El-Toni AM, Labis JP, Soylak M (2016) Synthesis and application of Fe_3O_4@ SiO_2@ TiO_2 for photocatalytic decomposition of organic matrix simultaneously with magnetic solid phase extraction of heavy metals prior to ICP-MS analysis. Talanta 154:539–547

34. Iqbal A, Ahamad T, Qais FA, Ahmad N, Shafi A, Ahmed AS, Srivastava S (2023) Proficient visible-light-driven photocatalytic and anti-biofilm activity of biosynthesized CeO2-graphene oxide nanocomposites. Mater Chem Phys 298:127397
35. Demir AY, Topkaya R, Baykal A (2013) Green synthesis of superparamagnetic Fe_3O_4 nanoparticles with maltose: its magnetic investigation. Polyhedron 28(65):282–287
36. Lu W, Shen Y, Xie A, Zhang W (2010) Green synthesis and characterization of superparamagnetic Fe_3O_4 nanoparticles. J Magn Magn Mater 322(13):1828–1833
37. Wang T, Ai S, Zhou Y, Luo Z, Dai C, Yang Y, Zhang J, Huang H, Luo S, Luo L (2018) Adsorption of agricultural wastewater contaminated with antibiotics, pesticides and toxic metals by functionalized magnetic nanoparticles. J Environ Chem Eng 6(5):6468–6478
38. Ahmad N, Bano D, Jabeen S, Ahmad N, Iqbal A, Anwer AH, Jeong C (2023) Insight into the adsorption thermodynamics, kinetics, and photocatalytic studies of polyaniline/SnS2 nanocomposite for dye removal. J Hazard Mater Adv 10:100321
39. Ahmad N, Sultana S, Sabir S, Khan MZ (2020) Exploring the visible light driven photocatalysis by reduced graphene oxide supported Ppy/CdS nanocomposites for the degradation of organic pollutants. J Photochem Photobiol A: Chem 386:112129
40. Ahmad N, Sultana S, Faisal SM, Ahmed A, Sabir S, Khan MZ (2019) Zinc oxide-decorated polypyrrole/chitosan bionanocomposites with enhanced photocatalytic, antibacterial and anticancer performance. RSC Adv 9(70):41135–41150
41. Cheng M, Zeng G, Huang D, Lai C, Xu P, Zhang C, Liu Y (2016) Hydroxyl radicals based advanced oxidation processes (AOPs) for remediation of soils contaminated with organic compounds: a review. Chem Eng J 15(284):582–598
42. Mi X, Wang T, Chen Y, Cao D, Ma N, Dai W (2025) Ingenious construction of a magnetic-recyclable photo-Fenton catalyst $ZnFe_2O_4$@ MIL-88A (Fe) and its adsorption-degradation activity toward levofloxacin. J Environ Sci 1(151):677–691
43. Ahmad A, Khan ZU, Sabahat S, Sun J, Shah NS, Khan ZU, Muhammad N, Mir S, Rahim A, Nadeem M, Khasim S (2024) Innovations in metal oxides-biochar nanoparticles for dye removal. Nano Struct Nano Objects 1(39):101269
44. Zhu Y, Fan W, Zhou T, Li X (2019) Removal of chelated heavy metals from aqueous solution: a review of current methods and mechanisms. Sci Total Environ 15(678):253–266
45. Tlili H, Elaoud A, Fuentes-García JA, Horchani-Naifer K, Ferhi M (2025) Magnetite nanoparticle-infused nanofibers composite: a green solution for industrial dye wastewater treatment. Emerg Mater 25:1–6
46. Almomani F, Bhosale R, Khraisheh M, Almomani T (2020) Heavy metal ions removal from industrial wastewater using magnetic nanoparticles (MNP). Appl Surf Sci 15(506):144924
47. Chen F, Xie S, Huang X, Qiu X (2017) Ionothermal synthesis of Fe_3O_4 magnetic nanoparticles as efficient heterogeneous Fenton-like catalysts for degradation of organic pollutants with H_2O_2. J Hazard Mater 15(322):152–162
48. Wang N, Zhu L, Wang D, Wang M, Lin Z, Tang H (2010) Sono-assisted preparation of highly-efficient peroxidase-like Fe_3O_4 magnetic nanoparticles for catalytic removal of organic pollutants with H_2O_2. Ultrason Sonochem 17(3):526–533
49. Wani AA, Shahadat M, Ali SW, Ahammad SZ, Uddin MK (2022) Recent advances and future perspectives of polymer-based magnetic nanomaterials for detection and removal of radionuclides: a review. J Mol Liq 1(365):119976
50. Aberdeen S, Cali E, Vandeperre L, Ryan MP (2023) Selective radionuclide and heavy metal sorption using functionalised magnetic nanoparticles for environmental remediation. J Mater Chem A 11(29):15855–15867
51. Zhan S, Yang Y, Shen Z, Shan J, Li Y, Yang S, Zhu D (2014) Efficient removal of pathogenic bacteria and viruses by multifunctional amine-modified magnetic nanoparticles. J Hazard Mater 15(274):115–123
52. Huang YF, Wang YF, Yan XP (2010) Amine-functionalized magnetic nanoparticles for rapid capture and removal of bacterial pathogens. Environ Sci Technol 44(20):7908–7913

Nanomaterials for the Removal of Metal Ions from Wastewater

Basree, Arif Ali, and Musheer Ahmad

Abstract While the fact that heavy metals are frequently occur in minute quantity in natural streams, the abundance of heavy metal ions are much dangerous for human survival and aquatic life at their very low concentrations. Nanotechnology and nanomaterials have put out the interest of experts as a long-term solution to water pollution. Researchers are working to create innovative, commercial nanomaterials for application in water/wastewater remediation. Nanomaterial adsorbents are very advantageous for removing the heavy metal ions because of their huge specific surface areas and improved active sites for pollutant adsorption. The chapter addresses nanomaterial adsorbents, including synthesis techniques, nanomaterial kinds, factors affecting nanomaterials adsorption, as well as future problems. Chemical Vapor Deposition was discovered to be the most effective in providing required properties among various synthesis processes. The optimal removal capacity of some nanomaterials such as graphene composites and cellulose-based nanomaterials has been reported1989 mg/g for Hg(II) and 1641 mg/g for Cd(II), respectively. Other factors are also played an important role (such as contact time, effect of pH, adsorbent dose, temperature, and ionic behavior of metal ions) to remove toxic metal ions in large amount from aqueous waste. Moreover, nanomaterials have been proven as a promising adsorbent for adsorption of heavy metals ions.

Keywords Nanomaterials · Adsorbents · Heavy metal ions · Effect of factors · Wastewater treatment

---

Basree · A. Ali · M. Ahmad (✉)
Department of Applied Chemistry, ZHCET, Faculty of Engineering and Technology, Aligarh Muslim University, Aligarh, U.P., India
e-mail: amusheer4@gmail.com

A. Ali
Department of Chemistry & Chemical Biology, Indian Institute of Technology (ISM), Dhanbad 826004, Jharkhand, India

S. A. Bhawani and M. Jawaid (eds.), *Nanomaterials for Separation of Hazardous Contaminants from Wastewater*, Smart Nanomaterials Technology,
https://doi.org/10.1007/978-981-95-6079-0_8

1 Introduction

Living sustainably requires a supply of potable water. Water contamination is a serious concern, globally. Industry development, such as the expansion of energy plants and mining sectors, as well as environmental disasters caused by climate change, has all contributed to the shocking increase in dangerous contaminants in the aquatic environment. Water contamination, along with an enhancement in demand owing to population development, leads to raised lack of supply to safe potable water in many parts of the world [1]. Common water pollutants include, inorganic, organic, microscopic, and biological contaminants, which can cause major health concerns. This is particularly true for wastewater containing heavy metal ions. The heavy metal ions are a serious environmental problem because of persistence in water without any complexation and other non-biodegradable for long time at very low concentrations. That will be a part of food chain, destroying aquatic life and causing major health risks [2]. Heavy metals possess atomic weights ranging from 63.5 to 200.6 and densities more than 5 g/cm^3. Many industries produce toxic sewage that contain heavy metals such as Cd, Hg, Cr, Ni, As, Co, Zn, Cu, and Pb [3]. Some industrial waste matter are dumped into the environment and watersheds untreated or improperly treated. Furthermore, the proliferation of insufficient landfill sites has exacerbated water pollution. Cadmium can induce kidney damage and renal disease. It is primarily produced by the alloying, smelting, and electroplating industries. In the similar way, copper, which is primarily derived from the electronic industry and device trash, may result in Wilson disease, liver damage, and insomnia. Nickel is mostly derived by the electroplating, mineral processing, and battery manufacturing industries. It has been linked to persistent lung cancer, dermatitis, and asthma. Lead is primarily derived by the battery, ore smelting, fertilizer, metal plating, and pesticide industries. It has the potential to induce brain damage, reproductive troubles, digestive issues, and anemia. Heavy metals frequently make their way into the food chain through plants and fish. Heavy metals dissolved in water are taken up by plants and accumulate at their roots before being translocated to edible sections such as fruits and vegetables [4].

In recent years, different types of wastewater treatment and water purification technologies have been emerged to remove the heavy metal ions from waste such as adsorption, chemical precipitation, coagulation, oxidation, reduction, ion exchange, membrane separation, and reverse osmosis, all these techniques could be utilized as a convincing solution to produce fresh water [5]. However, these methods are used to remove various contaminants from aqueous waste, but they have several limitations such as cost effectiveness, complexation process, limited capacity, and low recyclability). In this scenario, adsorption process is a one of the much efficient techniques due to their easy accessible, cost effective of their materials (that are easily available such as agricultural waste, chitosan, and fly as hand chitin), high recyclability, specific surface area, better durability, and so on.

Moreover, huge surface area, active sites, and immobilization of the fabricated nanomaterials are used to remove heavy metals from waste. Nanomaterial-based adsorbents possess several distinct properties, including nano size, huge surface area, high reactivity, high solution mobility, high mechanical properties, porosity, hydrophobicity, dispersibility, and hydrophobicity. Several nanomaterials have explored to develop good adsorbents to remove the toxic metal ions from wastewater, including biosorbents, graphene, activated carbon, and magnetic nanoparticles (Table 1).

2 Mechanism for Removal of Heavy Metals

We still don't know and are debating the interaction mechanisms that enable heavy metals to be eliminated from water-based solutions. The majority of current methods for eliminating these pollutants rely on photocatalytic degradation, sorption, or sorption-reduction [6]. To comprehend their potential environmental impact and offer alternate elimination pathways, it appears necessary to determine the physicochemical features of these pollutant and integrate current characterization (such as spectroscopy) with theoretical calculations at the molecular level. Sorption method is the much interesting and fundamental technique to absorb the heavy toxic metal ions by utilizing supramolecular chemical interactions. The functional carboxylate, hydroxyl, and other oxygenated functional groups are showed supramolecular features of fabricated sorbent materials. During this process, sorbate materials is used as a function of sorbed (adsorbed or absorbed) or sorbent that is known as sorption [7]. Hydrophobic nature, donor–acceptor, and supramolecular interactions (hydrogen bond interactions, and/or electrostatic attraction) play a promising role in mediating the formation of strong surface complexes that control this process. Higher oxidative metal ions can be immobilized and converted into low-oxidative metal ions by the sorption method. The initially reduction of higher oxidation state of metal ions produces denser, more readily perceptible particles or clusters. These sorption-reduction processes are well-known for converting Se^{4+} to Se^{2+} and Cr^{6+} to Cr^{3+}. Lower oxidations of metal ions are hence much more biocompatible in the natural phenomenon. It is believed that the mechanism guiding this method is based on the expectation that high valent metal ions will be significantly more mobile than low valent metal ions [8]. In conclusion, photocatalytic degradation is often used to remove many organic pollutants from solutions, including persistent organic pollutants, also called POPs, and low amounts of heavy metals. Based on photocatalytic processes, this strategy largely relies on the distribution of active site surfaces, visible-light absorption, mass transport mechanisms, and catalyst shape. The process of action for heavy metal ions may vary according to the metal ions and light sources used.

Table 1 Frequently used techniques for heavy metals removal and wastewater remediation

Method	Description	Advantages	Limitations
Coagulation and flocculation	Water is treated using positively charged substances. These compounds work by neutralizing the negative charge of pollutants and other dissolved particles in water. When metal ions/particles interact with the present chemicals and form a larger particles that are known as floc. Because of its weight, floc settle down during water treatment process, a process known as sedimentation	Comparatively accessible because alum is cheap source	Heavy metal removal is ineffective. It frequently works with a precipitation technique to promote efficient elimination Sludge generation
Precipitation	In this method, chemical agents have been used for the conversion of metal ions into metal hydroxide, metal sulfides, metal carbonate, or phosphate. Afterward, a filtration method is used to isolate the formed precipitate	This method is very easy and cheaply available with remarkable technique	Lower effectiveness to treat the polluted water with large amount heavy metals. utilizes massive amounts of precipitating agents This produces a tremendous volume of poisonous sludge Chemical stabilization is essential, as is effective precipitant removal
Ion exchange method	Ion exchange method involves in exchanging of ions (cations or anions) between exchange material and pollutants. For ion exchange medium; resins, synthetic organic compounds with large number of functional groups that capable to be exchange the cation or anion with own function. These materials can inorganic or either natural polymeric materials	This method is much utilized due to long-time reusability and quick kinetic process	Precipitation and sedimentation of metal occur slowly pH sensitive Low binding affinity as the presence of many free acids group Applicable only when low concentration of metal present Contamination of metal ions on ion exchange media

(continued)

Table 1 (continued)

Method	Description	Advantages	Limitations
Membrane technologies	Membrane process is utilized to remove trace particles and suspended from water via physical separation barrier (often a porous or semi permeable membrane)	High efficiency of elimination Easy to fabrication and operation There is no need for chemical additions Control and scaling, as well as space saving	Extreme fouling large amount of energy required because of inherent pressure-driven process Pretreatment could be required Secondary pollutant production is possible, and post-treatment is required
Electrochemical technologies	In an electrochemical cell there is use of anodic and cathodic reactions	Accessible, no need of chemicals, less sludge production, saves energy	Electric supply is very costly, not so effective as the metal ion concentration is low in aqueous solution, large hub of investment
Adsorption	This physical method is used to trap metal ions in adsorbent materials having adsorptive properties (i.e., highly porous, specific surface area, activeness of functional groups), via supramolecular (chemical or physical) interactions. In this process, adsorbent is recyclable or regenerated	Large range of adsorbents, cheaper, accessible	Time taking post treatment for maximum binding capacity, there should be hybridization in some adsorbents

3 Synthesis Techniques for Adsorbents

Two fundamental strategies for synthesising nanoparticles can be broadly categorized: top-down strategy and bottom-up strategy. The top-down method entails the reduction of bulk material to tiny particles [9]. The aforementioned methods are an advancement of those that were utilized to create particles that are lower than one. Top-down strategies rely on the elimination or division of bulk material to achieve the required attributes. The surface structure of this method is not ideal, which is a disadvantage [10]. For instance, lithographically produced nanowires may have several surface imperfections and structural flaws. Sol–gel manufacturing, electron beam lithography, atomic force microscopy, gas-phase condensation, and other synthesis methods are a few instances of those that make use of this strategy. Bottom-up methods are more economical than top-down methods because they produce

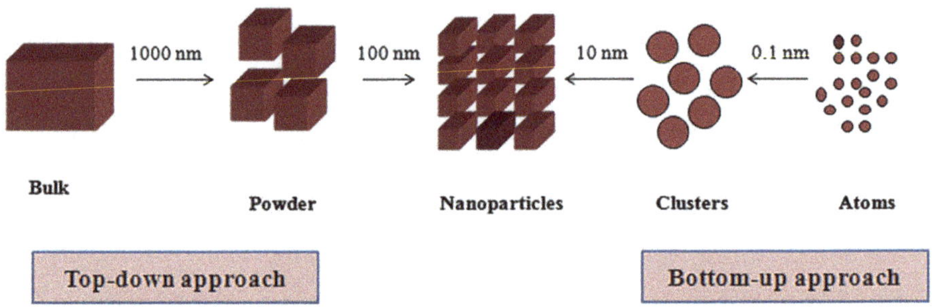

Fig. 1 Top-down and bottom-up strategies of nanoparticle synthesis

less waste. This technique alludes to synthesis procedures where the chemical is constructed from the ground up. It employs concepts of molecular self-assembly as well as recognition [11]. The bottom-up method is used to develop the uniform morphology of nanostructures with excellent crystallographic and surface structures. Examples include electrode positioning, template assisted sol–gel, hydrothermal synthesis, and other techniques. Because bottom-up offers various benefits over top-down, including fewer surface imperfections, better placing an order, and more uniform chemical composition, bottom-up is favored. The removal efficiency of nanomaterials is depend on the concentration of pH of solution, adsorbent dose, amount of toxic metal ions, contact time, and specific surface area [12] (Fig. 1).

4 Nanomaterials Used for Heavy Metal Removal

Removal of organic contaminants and heavy metal ions from wastewater, several nanomaterials as well as their composites are being utilized. Nanomaterials have uniform morphological size, specific surface area, good optical properties, better conductivity, fine particle size ratio, appropriate available adsorption sites, and magnetic properties. These types of material have shown better chemical and physical properties as compared of nanoparticles such as metals and their oxides, carbon derivatives and nanotubes, polymers composites, ceramics, and so on [13]. An ideal adsorbent should have following properties to remove heavy metal ions from contaminated water.

(i) Recyclable.
(ii) Safe for environment.
(iii) Should have high sorption efficiency as well as high selectivity at very low concentration.
(iv) Have that kind of surface by which adsorbed pollutant can be remove easily (Fig. 2).

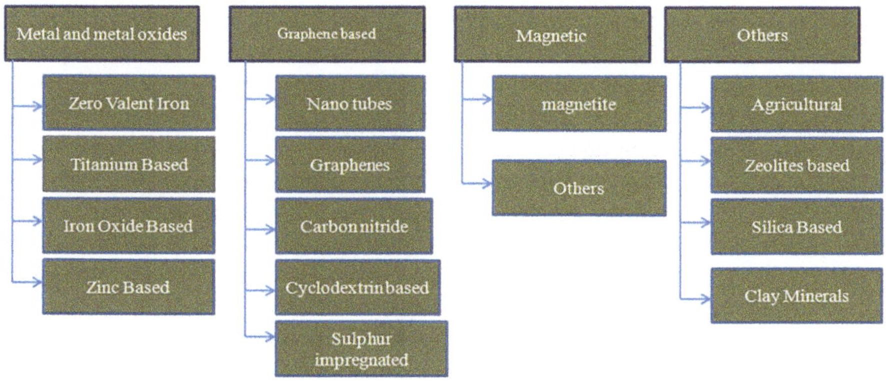

Fig. 2 Classification of nanoparticle materials

4.1 Metal Nanoparticles

Nowadays, nanoparticles of metal oxide play a substantial role in separation of heavy/toxic metal ions because of their beneficial well morphological advantages. Nanoparticles of iron, Gold, Ag, tin, and copper may be worked for elimination of metal ion-based pollutants in a two-step method, with the metal ions first adhering to the adsorbent's outer surface, followed by diffusion into the permeable walls of the particle via nano size corresponding. Their adaptability, merged with adsorption parameters such as contact's energy, a large number of active sites, oxidation state, and radius of metal ion as well as key contaminants prevalent in water, increases their appropriateness for application in heavy metal removal [14].

4.1.1 Zero-Valent Iron

Around the 1980s, zero-valent iron gained popularity for cleaning up groundwater in porous reactive barrier systems for removing pollutants such as perchlorates, chlorinated hydrocarbons, and nitro aromatics. Since then, Iron has been a potent remediation chemical due to its non-toxicity, abundantly availability (Fe is prevalent common element), and standard reduction potential (0.44 V) (which functions as a moderate RA) that often made up of a core shell [15]. The core shell is constructed of metallic iron (zero valent), and its outer shell forms on oxidation (mixed-valent: Fe^{2+} and Fe^{3+}).

$$Fe^0 + 2H^+ + O_2 \longrightarrow Fe^{2+} + H_2O_2$$
$$Fe^0 + H_2O_2 + 2H^+ \longrightarrow Fe^{2+} + 2H_2O$$
$$Fe^{2+} + H_2O_2 \longrightarrow Fe^{3+} + OH^- + OH^{\bullet}$$

Still, the main shortcomings of conventional zero-valent iron were the inability to locate the sources of pollution and the high building expenses. The development of nanoscale zerovalent has assisted in the method during the previous few decades. As a result of their tiny size, they are easily injected throughout the contaminated region using gravity's flow and pressure [16]. They stay suspended in the water under ambient conditions, cleaning it for a long time and exhibiting greater reactivity due to their larger surface areas. Numerous studies have even demonstrated the synergistic effects of bivalent iron-based nanoparticles (Ni, Pd, and Pt) in the breakdown of pollutants. A little doping of noble metals forms a thin layer on the surface of iron catalysis. The coated layer of these noble metal ion decreases the accumulation of corrosive products, while increasing the reaction kinetics. Among bivalent nanoparticles, the Iron-Palladium system is the most effective at removing polychlorinated biphenyl pollutants. The increased deterioration is due to a modification in the debromination phase caused by the presence of Palladium [17] (Fig. 3).

Emulsified zero-valent iron is another iron-based nanomaterial that is gaining popularity. It slows the breakdown of nano zero-valent iron and makes it easier to transfer in oil–water emulsion systems. They decompose chlorinated hydrocarbons by forming a thin layer of particles around the zero-valent iron and are surfactant stabilized [18]. Figure 3 illustrates the steps necessary to create a nano-emulsion gel.

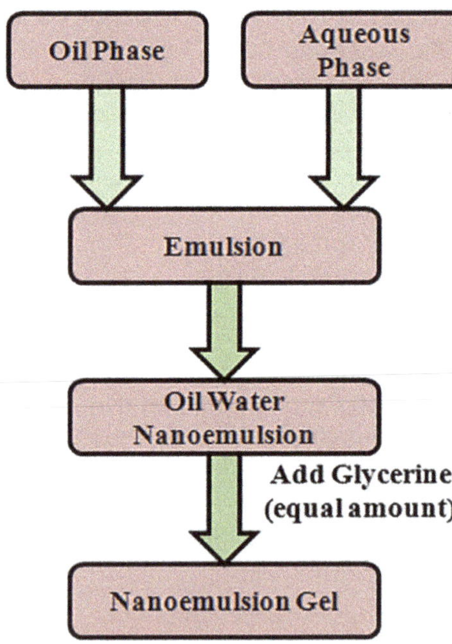

Fig. 3 Emulsion formation routes

4.1.2 Titanaate Nanomaterials

Thallium is a poisonous trace metal that is extremely hazardous to one's respiratory and gastrointestinal systems. Many nations have named it a top priority water contaminant since it is harmful even in small doses. According to recent instances in emerging countries such as China, there is an instant need to address this contaminant. Titanium nanomaterials with tubular/wire structures can effectively eliminate Thallium.

Apart from Thallium, it can be used to successfully remove a variety of other metals. Lead > Thallium > Copper > Cadmium > Zinc is the preferential order of adsorption for titanium.

Research using surface complexity models demonstrate that over a wide pH range, monodentate surface complexes based on titanium are superior to bidentate surface complexes for adsorbing arsenic [19]. Titanate nanoflowers produced under hydrothermal circumstances by the treatment of nano-anatase and sodium hydroxide demonstrate strong structural stability, resulting in a high reusability of the adsorbent. Even when interfering cations such as calcium and magnesium are present, they do not differ much. However, the main disadvantage is that these nanomaterials synthesized hydrothermally by treating TiO_2 and alkalis that are dependent on shape and crystal state [20]. They have comparable sorts of adsorption sites but vary in the positioning of functional groups throughout the linkage.

4.1.3 Iron Oxide-Based

Fe(II)is a most abundant of earth's metals, and it can exist in a different of oxide forms, such as hematite (alpha Fe_2O_3), magnetite (Fe_3O_4), and maghemite (gamma Fe_2O_3). It has recently demonstrated incredible sorption efficiencies owing to its enormous surface area, very permeable nature, and strong magnetic characteristics [20]. Colors, heavy metals, inorganic, and organic contaminants are frequently removed using this method size, surface flaws, and, especially, the ideal ratio of ferromagnetic and antiferromagnetic particles can all be used to control and govern the material's magnetic properties, resulting in a very stable structure. Like the majority of other metal oxides, it is prone to forming composites [9]. Inverse-spinal nickel ferrite ($NiFe_2O_4$) is one such extremely efficient composite. The functionality of composite has been shown an effective adsorbent properties due to the creation of substantial ferromagnetic properties like anti-parallel rotation between geometrical structures of iron and nickel (tetrahedral and octahedral).

As the industrial sector moves toward more environmentally friendly operations, procedures such as advanced oxidation processes (AOPs) utilizing semiconductor photocatalysis have grown in relevance. Conventional photocatalysts such as zirconium oxide, zinc oxide, titanium oxide, and others can only capture 5% solar energy [21]. To address the issue of insufficient use, researchers investigated photocatalysts with greater light capturing efficiency. One such substance is iron oxide, which, as hematite (alpha-Fe_2O_3), can absorb about 43% of incoming light in the visible spectrum. Its photocatalytic activity can be increased further by employing techniques such as doping with metallic or nonmetallic elements, morphology-controlled synthesis, composite creation, and so on. For example, Yu et al. used a hydrothermal technique to synthesize a BiOBr/Hametite composite and used it to select the degradation of the organic contaminant Rhodamine B. Synergistically, formation of p-n hetero-junctions between BiOBr and hametite has been shown a significant degradation improvement [22].

4.1.4 Zinc-Based

Rock salt, zincblende, and wurtzite are the three distinct crystal forms of zinc oxide that can be found naturally in the rare mineral zincite [23]. ZnO utilizes a four-step heterogeneous catalysis to remove organic pollutants.

(i) Production of charge carriers.
(ii) Trapping of charge carriers.
(iii) Recombination of charge carriers.
(iv) Photocatalytic degradation of organic pollutants.

The mechanism of photocatalysis is as given below [24].

Photocatalyst + hv	→	$e^- + h^+$
H_2O	→	$H^+ + OH^-$
$e^- + O_2$	→	O_2^{2-}
$h^+ + OH^-$	→	OH^*
h+ + OH-	→	OH^-
OH^* + R(pollutant)	→	OH^- + R*(oxidized)
$O_2^- + R$	→	O_2 + R**(reduced)
R*(reduced) or R**(oxidized)	→	(Final products) → $CO_2 + H_2O$

An optimal temperature of almost 600 °C is usually recommended for successful photo splitting of water. This is due to the sample's increased optical absorption and lower electric al resistivity at these temperatures [25]. These photocatalysts exhibit distinct excited states that are depend on the wavelength of irradiation, resulting in various reaction routes for different contaminants. This procedure also has the advantage of not requiring any secondary disposal methods, being environmentally benign, employing sustainable technology, and leaving no evidence of contaminants.

4.2 Magnetic Nanoparticles

Magnetic nanoparticles, differentiated by their distinct structure, size, and morphology, provide significant benefit in water treatment. They outperform traditional replacements and demonstrate advanced benefits by combining nanotechnology and magnetic separation advantages. They are toxic, easily dispersible, biocompatible, and chemically inert. However, due to their propensity to aggregate in aqueous solutions and severe oxidation, they are essentially impractical. On the other hand, surface alterations such complex formation, ligand addition, and functionalization using poly vinyl alcohol or polyvinylpyrrolidine boost oxidation stability and increase adsorptive power [26].

4.2.1 Magnetite (Fe_3O_4)

Unlike most other iron oxides, this is a type of iron ore containing both Fe^{2+} and Fe^{3+} ions. It features an energy gap of about 0.12 eV and n-type and p-type semiconductor characteristics. Because of its superior characteristics, it has seen significant use in radionuclide removal. According to research, a Schiff-based compound of nanorod-decorated silicon may be utilized to remove Pb(II) and U(VI) from contaminated water [14].

By surface functionalization of DMSA, superparamagnetic iron oxide (Fe_3O_4) nanoparticles may also be employed like extremely efficient adsorbents to detoxify hazardous compounds such as mercury, lead, and cadmium. For these heavy metals, this is attached to DMSA ligands, but for arsenic, it attached to iron oxide lattices. So, iron can be removed from the solution utilizing 1.2 T magnet. Furthermore, they have a high functionality of 1.85 mmol thiols/g for a surface area of about 120 m^2/g. Ferromagnetic nanoparticles with hydroxyl groups have been shown to be useful for dye removal at low temperatures [27].

4.2.2 Other Magnetic Materials

Magnetic nanocomposites are able to be divided into three categories.

(i) Organic–Inorganic Nanocomposites: Separated nanoparticles that have been finely disseminated in a polymer.
(ii) Core–Shell Inorganic Nanocomposites: These are created by combining two nanoscale objects into an individual hybrid material.
(iii) Self-assembled Colloidal Nanocomposites: The formation of ordered macroscopic structures from minuscule building blocks.

In one study, graphene oxide and chitosan particles has been used to create magnetic nanocomposites, which had a 76.94 mg/g capacity for adsorbing Pb^{2+} ions. Studies have demonstrated the efficacy of magnetic charcoal, chitosan gel particles, and magnetite that included chemically immobilized copper phthalocyaninedye in removing malachite green, polycyclic dyes, and organic colors from water that has been contaminated [28].

Polymethylacrylate (PMA), a functionalized magnetite nano material, can be used to remove oil from contaminated water. Superconducting magnetic separation can be used to recover it because it exhibits considerable ferro-magnetism and an outstanding capacity to absorb oil. In a study, magnetic nanocomposite $Fe_3O_4@SiO_2$

starch graft polyacrylic acid successfully removed the organic contaminant crystal violet dye. 85% of the dye was broken down, primarily because of how magnetically sensitive the hydrogels were [29]. The use of magnetic hexacyanoferrate polymers to recover Cs from low-level radioactive waste is standard procedure. By using Polyacrylonitrile as a binder, maximum separation was achieved with a contact period of 2 h and irrespective of pH.

4.3 Carbon-Based Nanomaterials

4.3.1 Carbon Nanotubes

Long, cylindrical nanostructures called carbon nanotubes are produced of sp^2 hybridized carbon atoms from rolled graphite sheets. Carbon nanotubes have outstanding chemical stability, mechanical and electrical capabilities, specific mesoporous nature, and enormous surface area that utilized in removing of heavy metals ions [30]. Because of their high porosity, they can be utilized to remove nonpolar chemicals as well as ionic and polar molecules. They are divided into two types: single wall nanotubes and multi wall nanotubes. While multiwalled nanotubes features nested graphite cylinders with diameters around 225 nm, single walled nanotubes have a rolling sheet with a diameter about 12 nm. In comparison to non-functionalized ones, which have an adsorption capacity of 2.49 mg/g, functionalized ones generate around 2.517 mg/g. This is due to significant chemical structural alteration, such as functionalization with a 3:1 vol. ratio of nitric acid and potassium permanganate. Though carbon nanotubes have been widely employed for Pb(II) separation, these environmentally acceptable adsorbents have a low selectivity. In contrast, microwave heating speeds up the reaction's kinetics and achieves up to 95% effectiveness in removal at the appropriate temperature and pressure levels that have a pH of 7.5–8.5 [31].

4.3.2 Graphene-Based

Because of the enormous surface area that can be used for adsorption, graphene oxide, a combination of various functional groups such carboxylic, hydroxyl, and epoxide functional groups, demonstrates excellent removal efficiency of heavy metal ions [32]. Because of its inert surface chemistry and π-π interaction, which avoids contact with contaminants and hence reduces environmental deterioration, this is greatly efficient in aqueous medium. Stabilizers to graft functional groups are connected to their surface to increase their functionality. Stabilizers are commonly used in industry to remove the pollutants via supramolecular interactions. Graphene oxides are modified using other functionalized materials like octadecyl amine and tetrazolyl derivatives, that enhanced the co-adsorb properties at minimal concentrations of heavy metal ions in the primary-ternary-quaternary system [33].

Graphene-based nano-spinal ferrites are typically coupled with bismuth or cobalt ferrites to generate adsorptive NC with extremely effective pores and low energy requirements. The adsorption capacity of these composite is depend on the pi-pi stacking, complexation of the inner surface, and electrostatic interactions, more interaction are involved more will be adsorption ability [34]. This process is occurred by functional groups on the surface of composites; that are actively protonated and deprotonated over a variable pH range. The carboxylate and hydroxyl groups are interchange on the surface of $CuFe_2O_4$ composites on varying the pH or below pint zero charge for Arsenic(III) and Arsenic(V) that has confirmed by X-ray photon electroscopy analysis. The influence of pH of these composites is altered their surface charge and actively controls the adsorption process. Due to better chemical stability over a wide variety of temperature and pressure, graphene derived nano-spinal composites could also be eliminate organic pollutants by photocatalytic degradation.

They might be used as inexpensive substitutes for conventional adsorbents due to their low price and easy method of retrieval by using an external magnetic field [35]. The estimated cost and possibilities for reuse/recycling of any adsorbent play a crucial role in upscaling its use. Furthermore, magnetic separation under non-sensitive environmental conditions, as well as desorption with HCl and HNO_3, might be employed to recover it.

4.3.3 Carbon Nitride-Based (g-C_3N_4)

In the last several years, carbon nitride-based nanomaterials have made substantial progress in the processes and mechanisms of heavy ion removal owing to their diverse structure and strong sorption capabilities [36]. They are generally made using the stiff template approach, increasing the density of occupied sites on the surface. Doping is other method of surface modification that is widely utilized in these nanomaterials to modify physical parameters such as outer electron density, specific surface area available for electron exchange, and water solubility [37]. The Lewis acid–base model is frequently used to rationalize the formation of strong complexes between carbon nitride and metal ions. By using electric stabilization and electrostatic repulsion, modifications such as polymer coatings with polydopamine (PDA), melamine-based dendrimer amine, and polyethyleneimine (PEI) are employed to resist multilayer aggregation [38].

4.3.4 Cyclodextrin-Based

Researchers have recently shown a strong interest in using cyclodextrins to improve the structural and functional properties of carbon nanomaterials. Cyclodextrin-modified graphene oxide nanostructures have made substantial contributions to domains such as water cleanup, plant growth regulation, medication administration, and so on [39]. These hybrid complexes are very acid stable and hence have a higher potential for pollutant removal from water. It was employed for the adsorption of deionized water-prepared Rhodamine dye. The recyclability was tested for six cycles utilizing the absorption–desorption method, and the removal capacity was found to be over 90% after six cycles.

CDMIPs—cyclodextrin-based imprinted polymeric adsorbents for perfluorinated compounds—were discovered to be useful due to their shape, surface tunability, and huge number of binding sites. By reducing the ratio of cross-linking agents and raising the degree of polymerization, it is possible to produce insoluble cyclodextrin polymers that are especially helpful for water treatment. Various investigations and experiments have confirmed that these polymers remove pollutants at the parts per trillion level, as opposed to activated carbons and zeolites, which work at the *ppm* level [40].

4.4 Other Nanoparticles

4.4.1 Agricultural Waste Materials

Recently, heavy metal ion removal has made use of affordable, readily available biodegradable wastes such as sunflower stalks, tea leaves, wool, coffee grounds, cotton, and coconut husks. A full comprehension of the functional groups associated for the binding is necessary to investigate how the metal attaches to the biomass. The major component of plant cell walls, where they are most frequently discovered, is cellulose. Its effectiveness as a stratum is strongly influenced by the degree of polymerization, the accessible surface area, and structural characteristics like crystallinity. Depending on the sample and measuring method, the amorphous and crystalline components' relative content varies. As opposed to cotton cellulose, wood cellulose is more amorphous. Cellulose-polyacrylonitrile and cellulose-polymethacrylonitrile, also known as cellulosic graft copolymers, are synthesised by mixing acrylonitrile and methacrylonitrile monomers in aqueous media with hibiscus cannabinus fibers in the presence of a coercion-toulene redox pair to remove zinc and chromium. The adsorption of Zn^{2+} ions is faster than Cr^{2+} ions at 29.85 °C under the same conditions [41].

4.4.2 Zeolites-Based

Zeolites are extremely effective at reducing organic pollutants. Zeolite-based membranes, especially those coated with narrow polyamide layers, are excellent for RO because of their excellent water permeability and salt rejection capacity. They have a sponge-like structure with numerous (micro or nano) pores, but they are rigid [42]. NAP1zeolites have shown promising results in the elimination of metals for instance zinc, nickel, copper, chromium, and cadmium.

The framework of zeolite materials, which are crystalline hydrated aluminosilicate minerals with multiple channels and pores varying in size from nano to micrometres, makes them extremely porous. Due to the development of nanoscale particles and the presence of nanopores, they are among the few material forms that are referred to be nanomaterials [43]. Tetrahedral SiO_4^- and AlO_4^- units make up the crystalline aluminosilicate material known as zeolites. A system of pores and channels that can accommodate small molecules, water, or cations. SiO_4^- and AlO_4^- units share O^{2-} atoms at the corners while Si^{4+} and Al^{3+} atoms are placed in the center, resulting in uniform intra crystalline voids and channels. The positive cations Na^+, Mg^{2+}, K^+, and Ca^{2+} are drawn to the net negative charge produced by Al^{3+} and may linger in the voids, counteracting the negative charge of the framework. Additionally, the zeolite's transparent 3D framework structure is exceedingly stable, and the weak ion and molecule bonds in the pores and gaps make it easier to extract these ions without damaging the zeolite structure [44]. According to their Si/Al molar ratio, zeolites can be chemically classified into three groups: minimal silica zeolites (aluminum-rich zeolites A and X) when Si/Al is lower than 2, intermediate silica zeolites when Si/Al is between 2 and 5, and high silica zeolites (zeolite beta, ZSM-5) when Si/Al is above 5. As a consequence, Si/Al ratio alterations affect the zeolite framework properties, giving zeolite compounds a wide range of industrial applications. For instance, in low silica zeolites, the framework aluminum with a molar ratio of Si/Al 1 has the most cation exchange sites to balance it, leading to the highest cation contents and exchange capacities. To boost heat and acid stability properties, bigger silicon content intermediate silica zeolites were needed, which made them essential for catalytic applications [45]. Less polar organic molecules only cooperate poorly with water and other polar molecules, while high silica zeolites can adsorb substances more effectively. Compared to low and intermediate silica zeolites, which have heterogeneous surfaces and an organophilic-hydrophobic selectivity, high silica zeolites.

4.4.3 Silica-Based

Owing to its particular properties, such as programmable surface features, controlled pore size, and large surface area, a novel category of metal oxide nanoadsorbents known as silica-based nanomaterials holds tremendous potential for metal ion adsorption. These nanoparticles have become an important metal oxide class that may be used to eliminate heavy metals from water. Additionally, nanosilica is a harmless and environment friendly adsorbent. When reacted with thiol and amino groups, these nanoadsorbents also consist improved metal selectivity and adsorption capabilities. Li et al. modified silica gel with nitrilotriacetic acid (NTA-silica gel) to effectively remove Pb^{2+}, Cd^{2+}, and Cu^{2+} from wastes [46]. With adsorptive values for Cd^{2+}, Cu^{2+}, and Pb^{2+} of 53.14, 63.5, and 76.22 mg/g, respectively, the synthesized material quickly absorbed metal ions. A further investigation found that the elimination of Cu^{2+} ions was enhanced by the use of silica nanospheres with amino-functionalization [3]. The ability to adsorb increased linearly with the number of amino groups. The removal of Pb, Ni, and Cd was compared using three distinct silica-based nanomaterials: nanosilica spheres (SNHS), amino-functionalized nanosilica gel (NH_2-NSG), and non-functionalized silica-based nanomaterials. In comparison to SNHS for functionalized nanosilica, the highest adsorption capabilities for Cd^{2+}, Ni^{2+}, and Pb^{2+} utilizing NH_2-SNHS became 40.73, 31.39, and 96.79 mg/g, respectively. Using 3-aminopropylethoxysilane (APTES)-assisted chemical etching (Ag-ACE) of silicon waste generated during the diamond-wire saw cutting procedure, nanoporous silicon (NPSi) hybrid compounds were created in a distinct study [47]. The APTES-NPSi showed highly-effective ability to adsorb for Cr^{6+} from aqueous solution which was considerably pH dependent, with the greatest ability to adsorb at 103.75 mg/g after 60 min at room temperature. The process for Cr^{6+} adsorption was the reduction of Cr^{6+} to Cr^{3+} via protonated amino groups. Following a maximum of five adsorption cycles, this functionalized nanoadsorbent continues to work effectively. Mesoporous silica works well as an adsorbent for contaminants in the gas phase due to its variable pore diameter and wide surface area. Surface changes including the presence of hydroxyl groups facilitate adsorption and accelerate related processes such as wetting Grafting functional groups onto porous walls is a well-known method for modifying and improving the effectiveness of these types of adsorbents [48]. The development of novel catalysts and adsorbents is possible as a result of this surface alteration. The presence of amine groups on silica surfaces can boost the elimination efficiency of carbon dioxide as well as hydrogen sulfide by up to 80%.

4.4.4 Clay Minerals

Composites such calcium polysulfides, kaolite, rectorites, as well as others are becoming more and more crucial in the removal of environmental pollutants in addition to modifiers, stabilizers, and inorganic clays. They assist in preventing the transport and buildup of pollutants. Due to qualities including greater stability, less agglomeration capability, and other features, zero-valent iron support inorganic clays are more effective than exposed zero-valent iron, according to numerous studies [18].

5 Factors Affecting Adsorption Process

The use of nanomaterials or functionalized nanomaterials as adsorbents to eliminate heavy metal pollutants from sewage water has grown dramatically in recent years. Because of relevant functional groups, such systems have unique traits for instance a great specific surface area and a stronger propensity binding heavy metals. Accessible nanomaterial-based adsorbents which have advanced significant in wastewater treatment have been classified according to the material used/functionalized for generation of the nanoadsorbent. The most often utilized metallic/metal oxide-based systems can be further classified on the bases of the oxidation states of the oxides (zerovalent, bimetallic) and crystalline characteristics of metal oxides (Al_2O_3 nanoparticles and nanosilica). Those characteristics often determine the capacity and mechanism of heavy metal adsorption from wastewater. Different factors that affect the adsorption process are given as below.

5.1 *Effect of pH*

There is evidence that pH modulates the adsorption process, based on mechanisms that necessitate the existence of functional moieties at the adsorbent's surface. The degree of protonation that the functional groups experience, giving increment of electrostatic repulsions between the heavy metal ions and the protonated groups, appears to be the important factor that seems to control the pH dependency (Fig. 4).

This is frequently instance with the systems which have lower pH with significant proton availability in that medium. Other process that may occur in minimal pH conditions is competitive adsorption of hydronium ions (H_3O^+) with cation metal species, resulting in ineffective adsorption. At greater pH, however, hydroxylated complexes of metal ions may create, restricting action by blocking surface moieties. The conclusion that a moderate pH range leads in optimal adsorption is intriguing. There is proof by a study that shows that raising the pH to 5.0 from 2.0 increases the adsorptive removal of Hg(II) ions by chitosan-alginate nanoparticles, afterward adsorption capacity falls down, implying the presence of an optimum range in which the adsorbent performs the best.

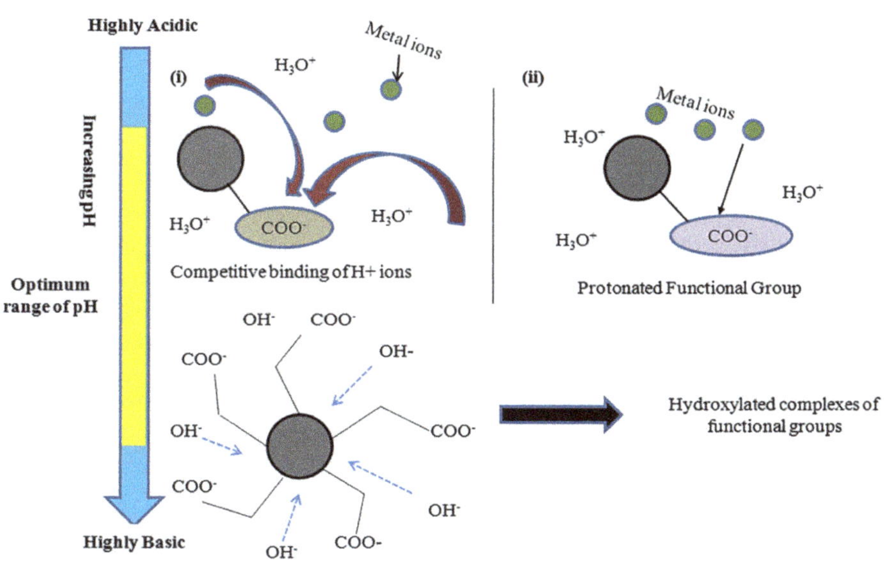

Fig. 4 Effect of pH on adsorption processes

5.2 *Effect of Contact Time*

The rate of removal is shown to improve gradually at first with an increase of interaction duration, and subsequently to slow down when the active sites are occupied. To supplement this pattern, a study looked at the adsorption of Pb(II) and Cd(II) ions on the surface of adsorbent. It was discovered that after the first 20 min, 100% and 91% of whole adsorption has completed, respectively, beyond which the procedure became unaffected by contact time because of the emergence of equilibrium [49, 50].

5.3 *Effect of Adsorbent Dose*

Adsorbent dosage enhance with an enhancement in the amount of accessible active sites for adsorption and thus makes better adsorption effectiveness. But a rise beyond a certain point causes aggregation of nanoparticles, which reduces surface area and active sites, lowering adsorption capacity. The chromium(VI) adsorption experiment revealed that at a fixed amount of the 7 g/L adsorbent, total adsorption process capacity declined. Similarly, the amount was raised from 4.0 to 20.0 g/L in another experiment for Cr(VI) elimination. Although maximum effectiveness was obtained after 120 min, adsorption seemed to diminish with increasing adsorbent quantity beyond a certain degree. This was projected to happen as a result of adsorbent particle overcrowding. Thus, increasing the adsorbent dosage is only advantageous till it raises active sites, at some point where efficiency begins to decline because of aggregation [51].

5.4 *Effect of Temperature*

The first rise in temperature is shown to diminish solution viscosity, resulting in improved diffusion ability thus leading to improved adsorption efficiency. Thermodynamically, the adsorption capacity falls down with increasing temperature in an exothermic process and enhances in an endothermic one. The adsorption of Hg(II) was tested to support this; it had discovered that the removal effectiveness raised first (from 10 to 30 °C, when it peaked), thereafter reduced. Some other arguments that prove this study are as follows [52]. At higher temperatures, enhanced mobility of metal ions causes the detachment of ions from the adsorbent, as do weakening electrostatic connections between the adsorbent and the metal ion.

5.5 *Effect of Ionic Strength and Initial Ion Concentration*

The quantity of metal ions accessible to adsorption becomes minimal at modest concentrations, resulting in decreased adsorption effectiveness. When it rises, so does the amount of ions accessible for adsorption, resulting in a faster adsorption rate. But, after the specific concentration is reached, total amount of saturated sites can no longer adsorb any more ions. As a result, the efficiency of adsorption reduces for the identical amount of adsorbent available. The influence of extra ions present in wastewater samples is described by the reliance of adsorbing capacity upon ionic strength of the aqueous solution. This is examined by varying ion concentrations such as Na^+ and Cl^-.

The Coulombic potentials are then measured to forecast the competition between the metal ions and other additional ions. Higher adsorbent molecule affinity for added ions rather than metal ions may reduce accessible active sites and be deleterious to the method.

6 Conclusion and Future Outlook

Heavy metal water pollution has been risen considerably in latest ages. As a result, technology, regulations, materials, and strategies should be created for addressing such a problem as well as get out of possibly disastrous consequences for human health as well as the environment.

Nanotechnology is a feasible option for dealing with heavy metal water pollution. Engineered nanoparticles' distinct physicochemical features enable them to effectively remove hazardous metal ions from contaminated water. The chapter has drawn attention to various synthesized nano-based materials that has employed and utilized from the adsorption method to separate heavy metal ions from sewage. Although the concentration of zeolites, magnetic, carbon materials, carbonaceous, magnetic and metal oxides as well as silica-based materials, polymer-based materials are very low but they demonstrated great removal effectiveness with various heavy metals via outstanding adsorption selectivity and efficiency. There are many significant characteristics that influence the progression of adsorption for example starting concentration, adsorbent dose, pH change, temperature change, and ionic strength. Optimizing the above variables may be crucial in eliminating hazardous metal ions quickly and establishing adsorption equilibrium. Few drawbacks have been identified with the usage of sorbents; under some cases, particles agglomerate and become unstable, reducing their removal effectiveness. Because of the tiny size, they require mechanical strength for separation of metal ions effectively. Long-term utilization of nanoparticles requires investigation. Second, more study is needed before nanomaterials may be commercialized to remediate industries or in situ heavy metal remediation. The operation and production costs of nanomaterials should be guided by green chemistry. Nanomaterial toxicity, health risks, and environmental impacts must all be considered. There is a scarcity of literature on real-time nanoparticle research. Nanomaterials for heavy metal treatment have persistent hurdles in terms of expenditure, fabrication, reusability, and separation. Furthermore, more research is needed to determine the negative effect nanomaterials might have on the environment and ecosystem.

Acknowledgements The Department of Applied Chemistry, ZHCET, Aligarh Muslim University, Aligarh, Uttar Pradesh, India-202002 for providing the laboratory facility. Basree thanks to the DST/INSPIRE Fellowship/2021/IF210635. A. Ali also thanks to the Institute Post-Doctoral fellowship (DAR/IPDF/CCB/53/2023), Indian Institute of Technology (ISM), Dhanbad. M. A. acknowledges the Start-up grants from the UGC, India.

References

1. Ahmad SZN, Wan Salleh WN, Ismail AF, Yusof N, Mohd Yusop MZ, Aziz F (2020) Adsorptive removal of heavy metal ions using graphene-based nanomaterials: toxicity, roles of functional groups and mechanisms. Chemosphere 248:126008. https://doi.org/10.1016/j.chemosphere.2020.126008
2. Ahmadi M, Hazrati Niari M, Kakavandi B (2017) Development of maghemite nanoparticles supported on cross-linked chitosan (γ-Fe_2O_3@CS) as a recoverable mesoporous magnetic composite for effective heavy metals removal. J Mol Liq 248:184–196. https://doi.org/10.1016/j.molliq.2017.10.014
3. Aigbe UO, Osibote OA (2020) A review of hexavalent chromium removal from aqueous solutions by sorption technique using nanomaterials. J Environ Chem Eng 8:104503. https://doi.org/10.1016/j.jece.2020.104503
4. Khudhair ZS, Zubaidi SL, Ortega-Martorell S, Al-Ansari N, Ethaib S, Hashim K (2022) A review of hybrid soft computing and data pre-processing techniques to forecast freshwater quality's parameters: current trends and future directions. Environments 9:85. https://doi.org/10.3390/environments9070085
5. Bolisetty S, Peydayesh M, Mezzenga R (2019) Sustainable technologies for water purification from heavy metals: review and analysis. Chem Soc Rev 48:463–487. https://doi.org/10.1039/C8CS00493E
6. Xiong C, Wang W, Tan F, Luo F, Chen J, Qiao X (2015) Investigation on the efficiency and mechanism of Cd(II) and Pb(II) removal from aqueous solutions using MgO nanoparticles. J Hazard Mater 299:664–674. https://doi.org/10.1016/j.jhazmat.2015.08.008
7. Ettre LS (1993) Nomenclature for chromatography (IUPAC Recommendations 1993). Pure Appl Chem 65:819–872. https://doi.org/10.1351/pac199365040819
8. Le AT, Pung S-Y, Sreekantan S, Matsuda A, Huynh DP (2019) Mechanisms of removal of heavy metal ions by ZnO particles. Heliyon. 5:e01440. https://doi.org/10.1016/j.heliyon.2019.e01440
9. Nizamuddin S, Siddiqui MTH, Mubarak NM, Baloch HA, Abdullah EC, Mazari SA, Griffin GJ, Srinivasan MP, Tanksale A (2019) Iron oxide nanomaterials for the removal of heavy metals and dyes from wastewater. In: Nanoscale materials in water purification. Elsevier, pp 447–472
10. Liu J, He S-H, Wang J-P (2020) High-yield gas-phase condensation synthesis of nanoparticles to enable a wide array of applications. ACS Appl Nano Materi 3:7942–7949. https://doi.org/10.1021/acsanm.0c01400
11. Hansen LT, Kühle A, Sørensen AH, Bohr J, Lindelof PE (1998) A technique for positioning nanoparticles using an atomic force microscope. Nanotechnology 9:337–342. https://doi.org/10.1088/0957-4484/9/4/006

12. Maitlo HA, Kim K-H, Kumar V, Kim S, Park J-W (2019) Nanomaterials-based treatment options for chromium in aqueous environments. Environ Int 130:104748. https://doi.org/10.1016/j.envint.2019.04.020
13. Singh S, Barick KC, Bahadur D (2013) Functional oxide nanomaterials and nanocomposites for the removal of heavy metals and dyes. Nanomater Nanotechnol 3:20. https://doi.org/10.5772/57237
14. Abdel Maksoud MIA, Elgarahy AM, Farrell C, Al-Muhtaseb AH, Rooney DW, Osman AI (2020) Insight on water remediation application using magnetic nanomaterials and biosorbents. Coord Chem Rev 403:213096. https://doi.org/10.1016/j.ccr.2019.213096
15. Li X, Elliott DW, Zhang W (2006) Zero-valent iron nanoparticles for abatement of environmental pollutants: materials and engineering aspects. Crit Rev Solid State Mater Sci 31:111–122. https://doi.org/10.1080/10408430601057611
16. Fu F, Dionysiou DD, Liu H (2014) The use of zero-valent iron for groundwater remediation and wastewater treatment: A review. J Hazard Mater 267:194–205. https://doi.org/10.1016/j.jhazmat.2013.12.062
17. Dave PN, Chopda LV (2014) Application of iron oxide nanomaterials for the removal of heavy metals. J Nanotechnol 2014:1–14. https://doi.org/10.1155/2014/398569
18. Galdames A, Ruiz-Rubio L, Orueta M, Sánchez-Arzalluz M, Vilas-Vilela JL (2020) Zero-valent iron nanoparticles for soil and groundwater remediation. Int J Environ Res Public Health 17:5817. https://doi.org/10.3390/ijerph17165817
19. Lata S, Samadder SR (2016) Removal of arsenic from water using nano adsorbents and challenges: a review. J Environ Manage 166:387–406. https://doi.org/10.1016/j.jenvman.2015.10.039
20. Xiang-Xue W, Xing L, Jia-Qi W, Hong-Tao Z (2019) Recent advances in carbon nitride-based nanomaterials for the removal of heavy metal ions from aqueous solution. J Inorganic Mater 436. https://doi.org/10.15541/jim20190436
21. Santhosh C, Velmurugan V, Jacob G, Jeong SK, Grace AN, Bhatnagar A (2016) Role of nanomaterials in water treatment applications: a review. Chem Eng J 306:1116–1137. https://doi.org/10.1016/j.cej.2016.08.053
22. Zheng Y, Li C, Meng X, Zhang Z (2016) A conjugated composite of α-Fe_2O_3 and BiOBr with enhanced visible-light-induced photocatalytic activity. J Mol Catal A Chem 421:16–28. https://doi.org/10.1016/j.molcata.2016.05.004
23. Baruah SK, Pal S, Dutta J (2012) Nanostructured zinc oxide for water treatment. Nanosci Nanotechnol Asia 2:90–102. https://doi.org/10.2174/2210681211202020090
24. Gunti S, Kumar A, Ram MK (2018) Nanostructured photocatalysis in the visible spectrum for the decontamination of air and water. Int Mater Rev 63:257–282. https://doi.org/10.1080/09506608.2017.1379264
25. Lee KM, Lai CW, Ngai KS, Juan JC (2016) Recent developments of zinc oxide based photocatalyst in water treatment technology: a review. Water Res 88:428–448. https://doi.org/10.1016/j.watres.2015.09.045
26. Wadhawan S, Jain A, Nayyar J, Mehta SK (2020) Role of nanomaterials as adsorbents in heavy metal ion removal from waste water: a review. J Water Process Eng 33:101038. https://doi.org/10.1016/j.jwpe.2019.101038
27. Ahmaruzzaman M (2019) Nano-materials: novel and promising adsorbents for water treatment. Asian J Water Environ Pollut 16:43–53. https://doi.org/10.3233/AJW190032
28. Liu Z, Yang H, Zhang H, Huang C, Li L (2012) Oil-field wastewater purification by magnetic separation technique using a novel magnetic nanoparticle. Cryogenics 52:699–703. https://doi.org/10.1016/j.cryogenics.2012.05.015
29. Pourjavadi A, Hosseini SH, Seidi F, Soleyman R (2012) Magnetic removal of crystal violet from aqueous solutions using polysaccharide-based magnetic nanocomposite hydrogels. Polym Int. https://doi.org/10.1002/pi.4389
30. Lal S, Singhal A, Kumari P (2020) Exploring carbonaceous nanomaterials for arsenic and chromium removal from wastewater. J Water Process Eng 36:101276. https://doi.org/10.1016/j.jwpe.2020.101276

31. Renu, Agarwal M, Singh K. Heavy metal removal from wastewater using various adsorbents: a review. Journal of Water Reuse and Desalination. 2017;7:387–419. https://doi.org/10.2166/wrd.2016.104.
32. Liu X, Ma R, Wang X, Ma Y, Yang Y, Zhuang L, Zhang S, Jehan R, Chen J, Wang X (2019) Graphene oxide-based materials for efficient removal of heavy metal ions from aqueous solution: A review. Environ Pollut 252:62–73. https://doi.org/10.1016/j.envpol.2019.05.050
33. Anjum M, Miandad R, Waqas M, Gehany F, Barakat MA (2019) Remediation of wastewater using various nano-materials. Arab J Chem 12:4897–4919. https://doi.org/10.1016/j.arabjc.2016.10.004
34. Sun W, Pan W, Wang F, Xu N (2015) Removal of Se(IV) and Se(VI) by MFe2O4 nanoparticles from aqueous solution. Chem Eng J 273:353–362. https://doi.org/10.1016/j.cej.2015.03.061
35. Lam C, James JT, McCluskey R, Arepalli S, Hunter RL (2006) A review of carbon nanotube toxicity and assessment of potential occupational and environmental health risks. Crit Rev Toxicol 36:189–217. https://doi.org/10.1080/10408440600570233
36. Gautam PK, Singh A, Misra K, Sahoo AK, Samanta SK (2019) Synthesis and applications of biogenic nanomaterials in drinking and wastewater treatment. J Environ Manage 231:734–748. https://doi.org/10.1016/j.jenvman.2018.10.104
37. Mauter MS, Elimelech M (2008) Environmental applications of carbon-based nanomaterials. Environ Sci Technol 42:5843–5859. https://doi.org/10.1021/es8006904
38. Han Q, Chen L, Li W, Zhou Z, Fang Z, Xu Z, Qian X (2018) Self-assembled three-dimensional double network graphene oxide/polyacrylic acid hybrid aerogel for removal of Cu^{2+} from aqueous solution. Environ Sci Pollut Res 25:34438–34447. https://doi.org/10.1007/s11356-018-3409-9
39. Li L, Fan L, Duan H, Wang X, Luo C (2014) Magnetically separable functionalized graphene oxide decorated with magnetic cyclodextrin as an excellent adsorbent for dye removal. RSC Adv 4:37114. https://doi.org/10.1039/C4RA06292B
40. Karoyo A, Wilson L (2015) Nano-sized cyclodextrin-based molecularly imprinted polymer adsorbents for perfluorinated compounds—a mini-review. Nanomaterials 5:981–1003. https://doi.org/10.3390/nano5020981
41. Demirbas A (2008) Heavy metal adsorption onto agro-based waste materials: a review. J Hazard Mater 157:220–229. https://doi.org/10.1016/j.jhazmat.2008.01.024
42. Mahmodi G, Zarrintaj P, Taghizadeh A, Taghizadeh M, Manouchehri S, Dangwal S, Ronte A, Ganjali MR, Ramsey JD, Kim S-J et al (2020) From microporous to mesoporous mineral frameworks: an alliance between zeolite and chitosan. Carbohyd Res 489:107930. https://doi.org/10.1016/j.carres.2020.107930
43. Mahdavi Far R, Van der Bruggen B, Verliefde A, Cornelissen E (2022) A review of zeolite materials used in membranes for water purification: history, applications, challenges and future trends. J Chem Technol Biotechnol 97:575–596. https://doi.org/10.1002/jctb.6963
44. Yoldi M, Fuentes-Ordoñez EG, Korili SA, Gil A (2019) Zeolite synthesis from industrial wastes. Micropor Mesopor Mater 287:183–191. https://doi.org/10.1016/j.micromeso.2019.06.009
45. Mahmoodi NM, Saffar-Dastgerdi MH (2019) Zeolite nanoparticle as a superior adsorbent with high capacity: synthesis, surface modification and pollutant adsorption ability from wastewater. Microchem J 145:74–83. https://doi.org/10.1016/j.microc.2018.10.018
46. Li Y, He J, Zhang K, Liu T, Hu Y, Chen X, Wang C, Huang X, Kong L, Liu J (2019) Super rapid removal of copper, cadmium and lead ions from water by NTA-silica gel. RSC Adv 9:397–407. https://doi.org/10.1039/C8RA08638A
47. Kotsyuda SS, Tomina V V., Zub YL, Furtat IM, Lebed AP, Vaclavikova M, Melnyk IV (2017) Bifunctional silica nanospheres with 3-aminopropyl and phenyl groups. Synthesis approach and prospects of their applications. Appl Surf Sci 420:782–791. https://doi.org/10.1016/j.apsusc.2017.05.150
48. Yen C-H, Lien H-L, Chung J-S, Yeh H-D (2017) Adsorption of precious metals in water by dendrimer modified magnetic nanoparticles. J Hazard Mater 322:215–222. https://doi.org/10.1016/j.jhazmat.2016.02.029

49. Alhan S, Nehra M, Dilbaghi N, Singhal NK, Kim K-H, Kumar S (2019) Potential use of ZnO@activated carbon nanocomposites for the adsorptive removal of Cd2+ ions in aqueous solutions. Environ Res 173:411–418. https://doi.org/10.1016/j.envres.2019.03.061
50. Hasanzadeh R, Moghadam PN, Bahri-Laleh N, Sillanpää M (2017) Effective removal of toxic metal ions from aqueous solutions: 2-Bifunctional magnetic nanocomposite base on novel reactive PGMA-MAn copolymer@Fe_3O_4 nanoparticles. J Colloid Interface Sci 490:727–746. https://doi.org/10.1016/j.jcis.2016.11.098
51. Wadhawan S, Jain A, Nayyar J, Mehta SK (2020) Role of nanomaterials as adsorbents in heavy metal ion removal from waste water: a review. J Water Process Eng. 33:101038. https://doi.org/10.1016/j.jwpe.2019.101038
52. Dubey R, Bajpai J, Bajpai AK (2016) Chitosan-alginate nanoparticles (CANPs) as potential nanosorbent for removal of Hg (II) ions. Environ Nanotechnol Monit Manage 6:32–44. https://doi.org/10.1016/j.enmm.2016.06.008

Nanomaterials for the Removal of Pharmaceutical Drugs from Wastewater

Shama Firdaus, Shaikh Arfa Akmal, and Aiman Ahmad

Abstract Nanoparticles' distinct properties have aided in the resolution of a wide range of environmental issues. The direct release of pharmaceutical drugs into water streams raises concerns about the potential hazardous impact on human health and aquatic ecosystems. Nanomaterials have proven to show an effective treatment of wastewater by eliminating and degrading the unused drugs released from households, industries, hospitals, and other locations. This chapter focuses on the features and benefits of nanomaterials in eliminating drug impurities through conventional and green innovative procedures. Furthermore, the drug removal mechanisms such as adsorption, photocatalytic degradation, and biodegradation, as well as the function of various nanomaterials based on the dimensionality (0D, 1D, 2D, 3D, and hybrid), are also discussed in detail. Thus, this chapter gives insightful information about the potential of nanomaterials for the elimination of pharmaceutical pollutants from wastewater, discussing the cutting-edge research articles for environmental remediation.

Keywords Nanomaterials · Pharmaceutical drugs · Wastewater treatment · Synthesis · Mechanism · Dimensionality · Environmental remediation

S. Firdaus · A. Ahmad (✉)
Department of Applied Chemistry, Zakir Hussain College of Engineering and Technology, Aligarh Muslim University, Aligarh, Uttar Pradesh, India
e-mail: aiman.ahmad1@gmail.com

S. Firdaus
e-mail: shamafirdaus27@gmail.com

S. A. Akmal
Department of Chemistry, Faculty of Science, Aligarh Muslim University, Aligarh, Uttar Pradesh, India

S. A. Bhawani and M. Jawaid (eds.), *Nanomaterials for Separation of Hazardous Contaminants from Wastewater*, Smart Nanomaterials Technology,
https://doi.org/10.1007/978-981-95-6079-0_9

1 Introduction

Pharmaceutical drug contamination in wastewater results primarily from the inappropriate disposal of industrial chemical waste, agricultural chemical-based herbicides and insecticides, and unused or expired medications [1]. Approximately 40–70% of these wastes come from human excretion of ingested medications, veterinary pharmaceuticals and hormones used in aquaculture and animal breeding such as antibiotics, anti-inflammatory drugs, steroids, etc. A significant amount of trash, including radioactive and chemotherapy drugs, disinfectants, and other pollutants, also finds its way into water bodies through sewage overflows, hospitals, healthcare facilities, and medical and research organizations [2]. Furthermore, around 20–30% of pharmaceutical waste is not properly eradicated in wastewater treatment plants, and metabolites and breakdown products produced during transformation contribute to pharmaceutical pollution when released into water bodies [3]. Pharmaceutical pollution in wastewater has become a significant environmental concern that requires immediate attention, reliable and innovative solutions, as it poses considerable threats to human health such as cancer, neurological disorder, endocrine disruption, reproductive problems and resistance to various medications, particularly in pregnant women and children. Also, exposure to low concentrations of pharmaceutical contaminants for long term leads to bioaccumulation. Additionally, it affects aquatic ecosystems by altering species populations by changing their physiology, disrupting nutrient cycles, increasing mortality rate and ultimately impacting the overall health of water bodies [4]. Various conventional approaches have been employed to mitigate pharmaceutical pollution, including activated sludge processing, filtration, sedimentation, chlorination, landfill disposal, incineration, chemical precipitation, and biological treatment. However, these methods have significant limitations, removing only 50–60% of drug contaminants while consuming considerable energy. Their inability to eliminate complex, dissolved drug molecules results in hazardous byproducts and air pollutants that can contaminate groundwater [5]. Furthermore, existing regulatory standards lack clarity, allowing pollutants to persist. These challenges underscore the urgent need for novel, innovative solutions to effectively reduce pharmaceutical contamination from wastewater and safeguard environmental and public health.

Recent studies on nanoparticles have shown that they are quite effective at removing pharmaceutical pollutants from wastewater. Nanomaterials range in size from 1 to 10 nm have distinct optical, thermal, mechanical, electrical, and biological properties with enhanced stability, durability, reusability, and regeneration capacity. Their enormous surface-area-to-volume ratio increases their reactivity and reaction rates with the pollutants, resulting in improved adsorption [6]. This makes it easier to convert and break down the complex chemicals found in pollutants. Under visible light, the surface of certain porous nanomaterials with photocatalytic properties can be conveniently functionalized and tailored for effective pollutant extraction. Their biocompatibility decreases toxicity, antibacterial characteristics reduce biofouling, and biodegradability allows for safe decomposition, which aids in environmental sustainability [7]. They can be easily self-assembled at a low cost and scaled up for large-scale industrial applications [6]. Because of these features, nanoparticles have shown to be a new technology that is extremely suited and flexible for the removal of simple to complex compounds present in pharmaceutical drugs found in diverse wastewater bodies, solving the restrictions of traditional technologies.

This chapter provides a thorough analysis of different types of nanomaterials and their properties based on their dimensionality (0D, 1D, 2D, 3D, and hybrid materials), that allows successful removal of pharmaceutical pollutants (antibiotics, hormones, and pesticides) in industrial, municipal, and agricultural wastewater through adsorption, photocatalysis, and biodegradation removal mechanisms. Focusing on recent advancements, the chapter explores novel high-performance nanomaterial systems, scalable conventional fabrication methods, and green innovative synthesis techniques, such as sol-gel, spray drying, chemical vapor deposition, hydrothermal/solvothermal, microwave, ultrasonication, and electrospinning synthesis. This insightful research advancement in expanding nanomaterial-based wastewater treatment methods establishes a foundation for developing innovative nanomaterial models and sensors using cost-effective synthesis techniques to promote sustainable pharmaceutical degradation solutions.

2 Synthesis Techniques of Nanomaterials

The synthesis of nanomaterials has advanced significantly since the middle of the nineteenth century, going through two different generations of methodologies: conventional and green innovative novel synthesis techniques. These are discussed below.

2.1 Conventional Approach

For many years, conventional methods of producing nanomaterials have been widely used, including a variety of physical, chemical, and thermal procedures that need high temperature, energy, and pressure such as sol-gel, precipitation, spray drying, ball milling, chemical vapor deposition, and so on. This approach remains essential for nanomaterial manufacturing due to its established track record and widespread adoption, despite its limitations, which include high costs, long processing times, hazardous chemicals, toxic byproducts, and limited control over nanomaterial features. Some of them are discussed below.

2.1.1 Sol-Gel

It is a versatile wet chemical process for the synthesis of nanomaterials by conversion of molecular precursors into nanoscale materials through a series of hydrolysis

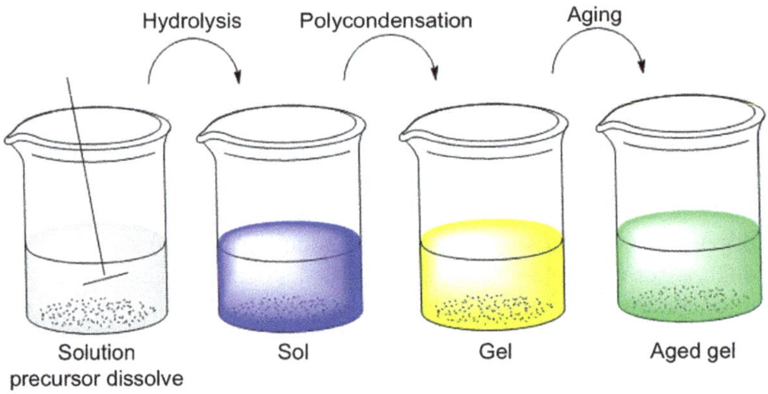

Fig. 1 Sol-gel graphical representation

and condensation reactions. As shown in Fig. 1, this process involves the formation of sol when precursor reactants metal alkoxides or chlorides dissolved in a solvent. The sol subsequently undergoes a condensation reaction to create metal–oxygen–metal bond, transitioning to the gel phase. The gel is then aged, making the network structure more consistent and stable. Finally, the gel is dried and calcined to remove organic residues and crystallize the nanoparticles [8]. For instance, sol-gel synthesis of wurtzite-structured ZnO nanoparticles with an ideal NaOH content successfully showed the photocatalytic degradation of pure and raw paracetamol under UV light. Because of the presence of zincate ions, the ZnO sample generated with lower NaOH concentrations proved to be a potent catalyst for the disintegration of paracetamol. Additionally, it showed reusability up to four times and eliminated 83% of paracetamol in 180 min when combined with H_2O_2 and UV light with a 0.0097 min^{-1} rate of reaction [9]. Another article reported the removal of carbamazepine (CBZ), an anticonvulsant drug, from wastewater through an adsorption mechanism by developing functionalized silica nanoparticles. In this, a strong hydrogen bond between CBZ and surface functional groups is formed. At room temperature and pH 9, with 15 min of contact time, 90% of CBZ removal was made possible due to the high surface area, uniform pore structure, and altered surface characteristics of silica nanoparticles [10]. Several other nanomaterials synthesized using this process, such as TiO_2, MgO, CuO, and Al_2O_3, have been effectively employed to remove ciprofloxacin, diclofenac, paracetamol, and other pharmaceuticals from wastewater. As a result, this method effectively captures and eliminates pharmaceutical contaminants from wastewater, showing promise for environmentally friendly and scalable water treatment options.

2.1.2 Spray Drying

In this technique, a liquid solution is rapidly dried using hot gas to produce a dry powder. In order to produce nanomaterials, precursor solutions are atomized into a stream of hot gas, which causes the solvent to evaporate immediately and generate dry nanoparticles [11]. The first step in this spray drying method involves forming a stable, homogeneous liquid solution or emulsion by combining chemicals for the required modifications. After that, the solution is put into a spray drier, where an atomizer turns it into precisely measured droplets. The solvent evaporates as these droplets go through a hot air chamber, leaving behind solid nanoparticles that are then collected after being separated from the air stream using a cyclone separator [12]. The graphical representation in Fig. 2 illustrates the process of spray drying method. Uniform, high-surface-area nanomaterials with regulated forms are produced by this ongoing process, allowing for customized characteristics for a range of uses. One article, for instance, described the manufacture of carbon-doped g-C_3N_4 granules using this method to remove tetracycline, which causes waterborne illnesses including cholera and typhoid in wastewater. By thermally breaking down a urea-thiourea combination and spray-granulating 2% polyvinyl alcohol (PVA) as a binder, followed by carbon doping, g-C_3N_4 nanosheets were created. This nanomaterial showed tetracycline removal by both photocatalysis and adsorption as they effectively eliminated 75% by adsorption in 60 min, and degraded photocatalytically > 95% in 90 min [13]. The photocatalytic oxidation of benzyl alcohol, a preservative used in many medicinal goods like ointments and injections, was reported in another paper. This study generated g-C_3N_4/TiO_2 granules by spray drying, which eliminated the need for high temperature or pressure by converting 94.5% of benzyl alcohol to benzaldehyde in 6 h at room temperature under UV–Vis irradiation with a yield of 81.2% [14]. It is simpler to remove benzaldehyde from wastewater since it is typically less toxic and more biodegradable than benzyl alcohol. More examples of drug removal from wastewater include spray-dried ZnO/TiO_2, Fe_3O_4/CNT, CuO/TiO_2, and so on. Therefore, it can be said that the spray drying technique of nanomaterials provides effective wastewater treatment by fine chemical synthesis, and organic pollutant degradation.

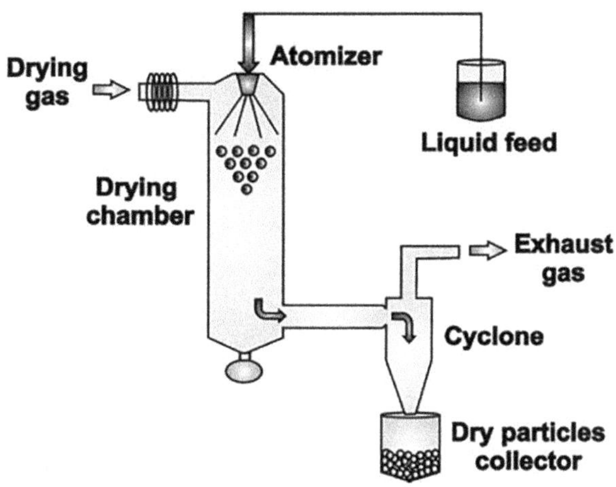

Fig. 2 Spray drying graphical representation

2.1.3 Chemical Vapor Deposition (CVD)

Nanomaterials, such as nanowires and nanotubes, can be readily synthesized by understanding the fundamental mechanism of the CVD method, which includes the kinetics that regulates rate of reaction, the thermodynamics that regulates chemical equilibrium and reaction kinetics, the diffusion that regulates precursor transport to the substrate, the nucleation that forms nuclei for nanomaterial growth, and the crystal growth theories that affect the material structure and morphology [15]. Figure 3 represents this process in which the precursor gases are broken down into reactive species in a heated vacuum chamber. These reactive species are then absorbed onto the substrate surface by chemical reactions that are controlled by temperature, pressure, and precursor flow rates, forming a tailored nanostructure. As the nucleation sites continue to be deposited, their growth and development ultimately result in the production of nanomaterials [16]. For example, many nanomaterials developed from CVD approach such as carbon nanotubes, graphene nanosheets, zinc oxide nanowires, molybdenum disulphide, titanium dioxide nanotubes [15] and so on effectively removed pharmaceutical drugs such as naproxen, ketoprofen, caffeine, tetracycline, sulfamethoxazole that are mostly not effectively degraded in conventional wastewater treatment process [17]. Despite its benefits, the inability to control the reaction results in non-uniform thickness since uniform deposition on large surfaces becomes challenging. Because of its toxicity, high expense, and post-deposition processing, which produces waste and pollutants, it is also more environmentally problematic.

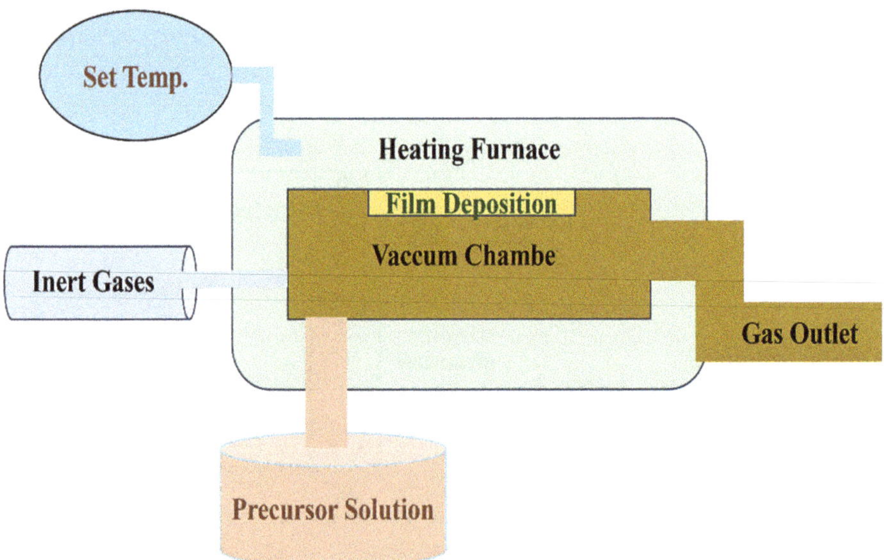

Fig. 3 Chemical vapor deposition graphical representation

2.2 *Innovative Green Approach*

Recent research on green and modern techniques for fabricating precised properties of nanomaterials from plants, microorganisms, and enzymes provides a sustainable and eco-friendly alternative to the traditional method. These techniques apply the principles of engineering, materials science, and nanotechnology to develop tailored and high-surface-area nanomaterials. Nanomaterials from this approach have transformed wastewater treatment and guaranteed the successful removal of pharmaceutical pollutants by demonstrating improved adsorption, degradation, and reusability for targeted drug removal. They also offer great efficiency, selectivity, and better water quality while reducing energy usage and costs. The versatility, sustainability, and affordability of advanced nanomaterials made with these state-of-the-art methods result in performance that is unparalleled. Few of these methods are discussed below:

2.2.1 Hydrothermal/Solvothermal

Temperature, pressure, and solvent characteristics and interactions influence the synthesis of high-crystalline pure nanomaterials in the hydrothermal/solvothermal process, which is a flexible and effective method based on the principles of thermodynamics and kinetics. Researchers have tailored the synthesis of different nanomaterials of specific size and shape, for the removal of pharmaceuticals from wastewater by modulating the reaction conditions [18]. This one-pot synthesis method involves making a solution of the reactant mixture and then heating it to the required temperature for a predetermined amount of time in a Teflon-lined autoclave. The autoclave is heated and then allowed to cool. To acquire the synthesized nanomaterials, the solution mixture inside the Teflon container is collected by centrifugation and filtration [19]. For instance, sewage and wastewater discharge include ciprofloxacin in its unmetabolized form. According to one article, ciprofloxacin was photocatalytically degraded by a novel magnetic nanophotocatalyst $ZnFe_2O_4$@CMC, which was prepared by hydrothermal process with carboxymethyl cellulose (CMC), a naturally occurring polymer produced from cellulose. At pH 7, when exposed to visible light for 100 min, this nanocomposite effectively eliminated ciprofloxacin around 87% from synthetic samples and 79% from real samples. Even after five cycles, this nanomaterial retained an 80% removal efficiency and shown good chemical stability and reusability [20]. According to another article, a ZnO-WO_3 nanocomposite synthesized via a hydrothermal process when exposed to visible light effectively photodegrades diclofenac, a non-steroidal anti-inflammatory medication that is a persistent pollutant in water bodies. About 95% of diclofenac was effectively broken down at pH 7 after 120 min, with a rate constant of 0.023 min^{-1}. They also demonstrated outstanding stability and reusability, retaining 85% of its initial activity after five cycles [21]. Thus, hydrothermally produced nanomaterials such as CuO, $Co(OH)_2$, and Fe_3O_4 can remove ibuprofen and acetaminophen from wastewater. As a result, these studies provide an opportunity to investigate new nanomaterial compositions for wastewater treatment methods in terms of scalability and cost-effectiveness for practical applications.

2.2.2 Microwave-Assisted Synthesis

This method heats chemical reactants using non-ionizing electromagnetic radiation in the 300 MHz–300 GHz range. It offers a rapid and energy-efficient method for producing a range of nanomaterials, such as metal nanoparticles (Au, Ag, Cu), semiconductors (CdS, ZnS), metal oxides (TiO_2, ZnO), and carbon-based materials (CNTs, graphene). Improved optical, electrical, and magnetic properties can be achieved with this technology in solvent-based and solvent-free procedures because of the faster reaction time, steady heating, higher yield, and controlled particle size [22]. Also, this technique is useful in the synthesis of complex nanostructures, such as core–shell nanoparticles and nanowires, by modifying nanomaterials for specific applications by the addition of functional groups or dopants. In this procedure, the precursor solution is prepared in a reaction vessel and subsequently placed in a microwave reactor. When microwave energy is applied for a specific amount of time, the temperature and pressure rise, causing dielectric heating and polarization of the molecules, which results in a chemical reaction. After the reaction mixture cools, the nanomaterials are separated by centrifugation, filtration, or precipitation [23]. For example, an article reported the successful synthesis of ZnAl-LDH/g-C_3N_4 nanomaterial composite by microwave-assisted technique that effectively decomposed the antibiotic ciprofloxacin. The use of a sustainable, metal-free, and environmentally friendly photocatalyst g-C_3N_4 makes composite synthesis greener. In about 10 min ZnAl-layered double hydroxide (LDH) and graphitic carbon nitride (g-C_3N_4) was synthesized under microwave irradiation. This nanomaterial degraded 97.5% of the ciprofloxacin when exposed to visible light in 120 min. The stability and reusability was also confirmed through five consecutive cycles [24]. Similarly, an article reported the synthesis of $ZnFe_2O_4$@methyl cellulose (ZFC@MC) nanomaterials using a microwave-assisted technique that effectively eliminated metronidazole, an antibiotic and antiprotozoal medication frequently found in wastewater, in just 10 min. Under ideal conditions, when ZFC@MC exposed to visible light it showed exceptional photocatalytic efficiency by degrading 92.65% metronidazole in synthetic samples and 71.12% metronidazole in real samples within 120 min. It performs 2.5 times better than bare $ZnFe_2O_4$ and has exceptional stability and reusability over six cycles without losing its catalytic efficacy [25]. More different types of nanomaterials such as ZnO nanorod, Fe_3O_4/SiO_2/TiO_2, $CoFe_2O_4$@methyl cellulose have been developed for the removal of many pharmaceutical drugs from wastewater. Therefore, these useful methods for eliminating pharmaceutical contaminants from wastewater using potentially biocompatible and magnetically recoverable nanoparticles offer effective and long-lasting water treatment.

2.2.3 Ultrasonication

This method generates high temperature and pressure in a solution of reactant mixture through the emission of high-frequency ultrasound radiation, which ranges from 20 kHz to 10 MHz. In this, the reactant solution is first prepared and then exposed to high-frequency ultrasonic waves that cause cavitation bubbles, fragmentation, and degradation of large particles into smaller ones. Extreme temperatures and pressures are generated when these bubbles burst with tremendous energy, promoting the nucleation and development of nanoparticles. Even particle distribution occurs due to the increased turbulence and vibration that enhances mixing and dispersion, causing the development and stabilization, ultimately producing the desired nanomaterial. The nanomaterial can be obtained by collecting, washing, and drying the resultant suspension [26]. For example, an article reported the synthesis of black Fe_3O_4 magnetic nanoparticles that were loaded by activated carbon produced from the olive kernel via ultrasonication was effectively used for adsorption of amoxycillin, an antibiotic generally found in wastewater. In this experiment, at pH 6 under optimal conditions with a contact time of 90 min, absorbance dose of 0.5 g/L nanomaterial showed 97.6% removal of 200 mg/L amoxycillin concentration. The adsorption was monolayer, homogeneous, and uniform [27]. Another article described the use of Hibiscus rosa-sinensis extract as a reducing and stabilizing agent for the synthesis of zero-valent copper (Cu^0) nanoparticles in order to degrade the drug 5-fluorouracil and lovastatin. In this experiment, the ultrasound aids in the controlled and uniform formation of amphoteric Cu^0 nanoparticles in the presence of Hibiscus rosa-sinensis extract, which increases the crystallinity with controlled growth of Cu^0. The combination of Cu^0 nanoparticles and ultrasound/UV light produces reactive radicals •OH and O2• that efficiently degrade 91.31% of fluoxetine, a basic pollutant, and 93.2% of lovastatin, an acidic pollutant. The amphoteric nature of Cu^0 permits interaction with both acidic and basic pollutants [28]. Numerous nanomaterials such as zinc oxide, silver, TiO_2@SWCNT have been synthesized for effective removal of sulfathiazole, levofloxacin, steroids, etc. by ultrasonication method. Thus, from the experiment it can be concluded that this technique offers fast, versatile, energy-efficient removal of the drugs from wastewater.

2.2.4 Electrospinning

This method has transformed the synthesis of nanomaterials and offers a plethora of novel alternatives based on the fundamentals of fluid dynamics and electrostatic forces. Electrical forces cause the precursor reactant solution to change into continuous Taylor cones and jet solidifying fibers when a high voltage is applied. As the solution leaves the cone, evaporation or solvent removal causes it to solidify into a nanofiber. These nanomaterials are perfect for a wide range of applications because of their high surface-area-to-volume ratio and tailored properties with large-scale production and cost-efficiency. The resultant nanomaterials have better mechanical, thermal, electrical, and optical qualities, making them appropriate for drug removal from wastewater due to their increased conductivity, chemical reactivity, and selectivity. For instance, bisphenol can have a serious negative impact on environmental health when it interacts with drugs in an aqueous environment. According to a recent publication, cellulose nanofibers modified with β-cyclodextrin (β-CD) were synthesized using the electrospinning process to generate a porous nanofibrous membrane (CA-P-CDP) that effectively removes bisphenol traces. This nanomembrane demonstrated effective absorption of bisphenol(A) 90.37%, bisphenol(S) 88.52%, and bisphenol(F) 87.25% in 15 min at pH 7 and room temperature. This result is due to the hydrophobic, hydrogen bonding, and π-π stacking interactions of the nanomembrane [29]. According to a different article, carbamazepine and non-steroidal anti-inflammatory drugs (ibuprofen, fenoprofen, ketoprofen, and diclofenac) were effectively removed from wastewater by the fabrication of nanofibers from a biopolymer composite composed of PVA and the water-soluble protein of moringa seeds. Under ideal circumstances, drug removal in the standard solution ranged from 84.5 to 97.4% but in real wastewater the removal of drugs ranged from 77.8 to 96.1% for the effluent and 70.54 to 92.5% for the influent [30]. These results suggest that the interaction between specific drugs and heterogeneous surfaces of nanofibers is regulated by multilayer physicochemical adsorption, concluding that electrospun nanofibers remove pharmaceutical drugs from wastewater easily, affordably, and ecologically. Some more different nanomaterial such as lignin, TiO_2, polymer/graphitic carbon nitride, chitosan/PVA nanofibers have been developed for effective removal of veterinary drugs, sulphonamides and various acidic and basic drug pollutants found in wastewater. The electrospinning technique has become a popular option for synthesizing nanomaterials because of its affordability, versatility, and ease of use, despite the fact that there are still issues to be addressed, such as managing fiber shape, increasing production volume, and combining electrospinning with other nanotechnologies.

3 Different Types of Mechanism for the Removal of Drugs from Wastewater

Several chemical and physical properties of the drugs and the nanomaterials play an important role in the removal of drugs from wastewater using nanoparticles. The nanomaterial properties, including the small size (nm), provide a large surface area for enhanced adsorption and reactivity, and the targeted drug removal is improved by the tailored pore size and shape with the uniform distribution of particles. The composition of nanomaterials affects stability and reactivity, whereas the surface chemistry and the zeta potential control chemical and electrostatic interactions between pharmaceuticals and nanomaterials, respectively [31]. Moreover, the pharmaceutical properties such as electrostatic interactions, hydrophobicity, and molecular weight, which cause diffusion, adsorption, and desorption with nanomaterials, also play a vital role in wastewater treatment [32]. The drug removal processes of various nanomaterials with particular features, such as adsorption, degradation, and separation, are discussed below.

3.1 Adsorption

Drug elimination processes include size exclusion, electrostatic attraction, and chemical and physical adsorption. During the adsorption process, the first step involves the diffusion process in which the concentration gradients and electrostatic forces cause the pharmaceutical drugs to interact with the surface of nanomaterials through weak intermolecular forces such as hydrogen bonds, van der Waals, and π–π interactions generating a thin layer. Subsequently, reversible, non-specific interactions and covalent bonds between the functional groups of drugs and nanomaterials, respectively, promote the physical and chemical adsorption. Drugs are trapped within nanomaterial pores due to electrostatic attraction and size exclusion. Equilibrium occurs when the maximum amount of the drugs is removed. Further, these drugs from nanomaterials can be desorbed after the adsorption process and the nanomaterial can be further reused for further wastewater treatment [33]. For example, an effective study reported the spontaneous, endothermic, and physical removal of non-steroidal anti-inflammatory drugs (NSAIDs) by the green-synthesized copper nanoparticles (Cu NPs). About 74.4, 86.9, and 91.4% of Ibuprofen, Naproxen, and Diclofenac were successfully extracted from the wastewater by Cu NPs within 60 min at ambient temperature and pH 4.5. For diclofenac, ibuprofen, and naproxen, the maximal monolayer adsorption capacities were determined to be 36.0, 33.9, and 33.9 mg/g, respectively [34]. Likewise, graphene oxide (GO) nanosheets, zinc oxide (ZnO) nanoparticles, and their combination were developed for levofloxacin adsorption. ZnO nanoparticles removed 83.4% of the levofloxacin at pH 6 in 30 min, GO nanoparticles removed 91.2% at pH 5 in 20 min, and the ZnO/GO (1:1 ratio) combination removed 97.5%. The combination showed a synergistic impact, while kinetic study showed pseudo-second-order adsorption for ZnO and GO. Even after five cycles, the reusability analysis showed that the combined ZnO-GO system had an 85% removal efficiency, making it far more effective than the separate nanocomponents [35]. Several other nanomaterials like Fe_2O_3, CNTs, MWCNT–TiO_2 composite have shown effective adsorption of acetaminophen, ibuprofen, sulfathiazole and many other drugs found in wastewater. Thus, from the above experiment and examples the adsorption method is known to be the most widely used mechanism for easy and sustainable wastewater treatment.

3.2 Photocatalytic Degradation

This mechanism employs light energy for the breakdown of pharmaceutical pollutants in wastewater. The nanomaterial absorbs light, causing electrons to be energized and migrate from the valence band to the conduction band, generating electron–hole pairs. The moderate band gap of nanomaterial separates electrons and holes, preventing charge recombination. Reactive oxygen species (ROS) are formed when the holes combine with water forming hydroxyl radicals and the electrons combine with oxygen forming superoxide radicals. Pharmaceutical contaminants are subsequently oxidized and dissolved by these ROS, resulting in the formation of smaller, nontoxic molecules like CO_2 and H_2O [36]. For instance, a Fe-doped TiO_2 nanomaterial with a band gap energy of 3.2–2.8 eV was synthesized using the surface impregnation method for the photocatalytic disintegration of carbendazim and propiconazole, which are categorized as veterinary pharmaceuticals (VPs), pesticide residues, or endocrine-disrupting chemicals typically found in wastewater. Different Fe wt% (1, 2, 3, and 4%) were loaded into TiO_2, but the 2 wt% of Fe loading resulted in 98.5–92% disintegration of carbendazim and propiconazole, respectively, under visible light conditions [37]. For the photocatalytic degradation of amoxicillin (AMC) and diclofenac (DCF) under visible light, another study described the synthesis of nanomaterials by doping nanoscale silver ferrite ($AgFeO_2$) on graphene (G) and $Cu_2(BTC)_3$ metal–organic framework (MOF) forming three distinct types of nanomaterials: $AgFeO_2$/G, $AgFeO_2/Cu_2(BTC)_3$, and $AgFeO_2$/G@$Cu_2(BTC)_3$ using a solvothermal technique. Out of these three nanomaterials, the tertiary $AgFeO_2$/G@$Cu_2(BTC)_3$ had graphene as a solid electron mediator/acceptor, and due to this, it demonstrated an enhanced charge transfer mechanism from the conduction band of $Cu_2(BTC)_3$ to the valence band of $AgFeO_2$, causing an improved degradation of drug molecules up to 97% on the catalyst surfaces within 150 min with stability up to four cycles [38]. Similarly, different nanomaterials such as carbon-based nanomaterials, $TiO_2/NaAlO_2$, Bi-modified titanate nanomaterials have been studied for photocatalytic degradation of amoxicillin, acetaminophen, aspirin, and so on and the study still continues. Thus, from the discussed examples efficient treatment of pharmaceutical wastewater under light (visible, UV, etc.) by photocatalytic mechanism offers a promising solution for environmental remediation.

3.3 Biodegradation

This method breaks down pharmaceutical pollutants from wastewater by utilizing the synergistic mechanism of microorganisms and nanomaterials. Microbes and nanomaterials are mutually activated by electrostatic forces, extrinsic polymeric compounds, and altered gene expression. In wastewater, microbes get attached to the nanomaterial by the formation of the biofilm which increases with time and activate each other. Microorganisms stimulate nanoparticles by dissolving the coating, producing ROS for oxidative degradation through photocatalysis, chemical reactions, and surface interactions, catalyzing redox reactions, and producing biosurfactants for dispersion. In contrast, nanoparticles stimulate microorganisms by providing attachment surfaces, generating ROS as a byproduct of cellular respiration and enzymatic reactions, and supplying nutrients. Finally, the enhanced ROS degrades the pharmaceutical pollutants into simpler molecules like CO_2, H_2O, and biomass [39]. A study was carried out in which the fungus Pleurotus ostreatus (P. ostreatus) was grown in order to produce laccase enzyme, and titania nanoparticles (TiO_2) were manufactured using the sol-gel method. In order to degrade the drug pollutants carbamazepine and bisphenol (A), the laccase enzyme was isolated and directly immobilized on TiO_2 NPs by physisorption at pH 5.5, 25 °C, and incubated for 2–24 h. 90% of the bisphenol A was removed in six hours, and when bisphenol A was present, the removal of carbamazepine increased from 10 to 40%. The breakdown of carbamazepine alone was only 10% because of its high redox potential, which prevented its oxidation from laccase enzyme [40]. To degrade diclofenac, another study described immobilizing the laccase enzyme on the surface of a composite nanofibrous membrane made of polyvinyl alcohol (PVA), chitosan (CS), and multi-walled carbon nanotubes (MWNTs). Composite membrane's porosity, large surface area, and permeability improves the stability, activity retention (76.7%), and enhances the enzyme loading up to 907 mg/g. Within six hours, diclofenac was completely removed from wastewater at pH 4 and 50 °C, and the nanocomposite was reusable for up to seven cycles with 62.7% efficiency [41]. Designing nanomaterials with effective biodegradation can help us cut pollution and advance sustainability. To provide standardized procedures for assessing biodegradation, additional investigation is required for a safer toxic-free water.

4 Recent Development in Dimensionality-Driven Nanomaterials for Pharmaceutical Drug Removal from Wastewater

Nanoparticles, categorized as 0D, 1D, 2D, or 3D materials, possess at least one dimension under 100 nm. This nanoscale size endows them with unique properties, particularly a significantly higher surface-area-to-volume ratio than conventional micrometer-scale sorbents. As a result, nanomaterials often exhibit superior adsorption capacities, improved removal efficiencies, and accelerated reaction kinetics. These attributes make them highly effective for diverse applications, including environmental remediation, catalysis, and drug delivery. In recent years, substantial scientific efforts have focused on designing and synthesizing innovative nanomaterials tailored to address specific challenges across various fields, especially for water purification, offering cutting-edge solution.

4.1 Different Types of Nanomaterial

4.1.1 0D Nanomaterial

Carbon dots (CDs), a class of zero-dimensional (0D) nanomaterials, have gained significant attention in recent years due to their excellent properties, including high surface area, tunable optical features, and the ability to generate reactive oxygen species (ROS) under light irradiation. These properties make CDs highly effective in environmental remediation, particularly for removing pharmaceutical contaminants from water. As 0D nanomaterials, CDs are small particles with all dimensions less than 100 nm, allowing them to exhibit high surface-to-volume ratios that enhance their interactions with pollutants. Carbon dots are often integrated into hybrid systems to boost their performance in photocatalytic applications. For example, in a study by Sendai et al. [42], CDs were combined with titanium dioxide (TiO_2), extending its light absorption range and enhancing its photocatalytic properties. The CD-TiO_2 hybrid demonstrated over 90% tetracycline degradation under visible light, showing the potential of CDs in removing antibiotic pollutants. Similarly, Yang et al. [43] explored a CD-graphitic carbon nitride (g-C_3N_4) nanocomposite for the degradation of ciprofloxacin. This system achieved a remarkable 95% ciprofloxacin removal, further proving the enhanced photocatalytic efficiency of CDs under visible light. Zhang et al. [44] developed a CD-BiOCl nanocomposite for the degradation of sulfamethoxazole, where the CDs significantly improved the photocatalytic performance, resulting in 90% removal of the pharmaceutical contaminant. In addition to photocatalysis, carbon dots are also effective in adsorption applications due to their high surface area, rich functional groups, and the ability to facilitate strong interactions with pharmaceutical contaminants. For instance, Shafi et al. [44] demonstrated the use of nitrogen-doped carbon dots (N-CDs) for the adsorption of ibuprofen. The N-CDs exhibited over 90% adsorption efficiency through hydrogen bonding and π–π interactions, showcasing their potential for removing non-steroidal anti-inflammatory drugs (NSAIDs) from wastewater. Lin et al. [45] investigated amino-functionalized carbon dots (A-CDs) to remove diclofenac, where the amino groups on the A-CDs enhanced the adsorption efficiency by enabling specific interactions with diclofenac molecules. Furthermore, Cao et al. [46] developed a carbon dot-magnetic Fe_3O_4 composite for tetracycline adsorption. This composite demonstrated high adsorption capacity, and its magnetic properties allow for easy recovery, making it a promising candidate for efficiently removing pharmaceutical contaminants. The unique combination of high surface area, functional groups, and tunable properties makes carbon dots versatile for photocatalytic and adsorption applications in pharmaceutical waste removal. Their ability to degrade and adsorb various pharmaceutical pollutants, such as antibiotics and NSAIDs, presents an effective strategy for addressing pharmaceutical contamination in aquatic environments. Integrating carbon dots with other nanomaterials further enhances their efficiency, making them a promising tool in environmental remediation.

4.1.2 1D Nanomaterial

One-dimensional (1D) nanomaterials, characterized by their elongated structures with one dimension much longer than the other two (typically nanowires, nanotubes, or nanorods), have emerged as efficient candidates for environmental remediation, including the removal of pharmaceutical contaminants from water. These materials, such as carbon nanotubes (CNTs), metal oxide nanowires, and titanium dioxide (TiO_2) nanorods, exhibit unique properties like high surface area, tunable aspect ratios, and enhanced charge transport, making them suitable for both photocatalytic and adsorption applications in wastewater treatment. In photocatalytic applications, 1D nanomaterials are highly effective in degrading pharmaceutical pollutants, mainly due to their high surface-to-volume ratios and the ability to promote charge separation. For instance, TiO_2 nanorods have been widely studied for their photocatalytic performance under UV and visible light. In a study by Guo et al. [47], TiO_2 nanorods were used to degrade tetracycline in wastewater, achieving over 80% degradation in 90 min. The high surface area of the nanorods allowed for effective pollutant adsorption, while their unique structure facilitated efficient charge separation, boosting the photocatalytic degradation process. Similarly, carbon nanotubes (CNTs), when combined with other photocatalytic materials, have been shown to enhance the degradation of pharmaceutical contaminants. For example, a CNT-TiO_2 composite demonstrated enhanced photocatalytic activity for the removal of ciprofloxacin under UV light, achieving 85% degradation in 60 min Hassani et al. [48]. 1D nanomaterials also excel in adsorption applications due to their high surface area, well-defined structures, and the presence of functional groups that enhance their ability to interact with pharmaceutical pollutants. For example, TiO_2 nanotubes, due to their large surface area and high porosity, have been studied for their capacity to adsorb and remove pharmaceutical contaminants. In a study by Valadez-Renteria et al. [49], TiO_2 nanotubes were used to adsorb diclofenac from water, with the material showing an adsorption efficiency of more than 90%. The nanotubes' ability to adsorb contaminants was attributed to their surface hydroxyl groups and porous structure, allowing easy interaction with the pharmaceutical molecules. Similarly, CNTs have been utilized for the adsorption of various drugs, including ibuprofen and acetaminophen. In a study by Oyetade et al. [50], functionalized CNTs exhibited over 85% adsorption efficiency for ibuprofen, demonstrating their effectiveness in removing NSAIDs from water. The unique structural advantages of 1D nanomaterials, such as their high surface area, enhanced charge transfer, and tunable surface chemistry, make them highly effective in photocatalytic and adsorption applications for removing pharmaceutical pollutants. Their ability to degrade and adsorb various pharmaceutical contaminants, including antibiotics and NSAIDs, underscores their potential as efficient and sustainable materials for environmental cleanup. Developing hybrid systems and integrating 1D nanomaterials with other nanostructures further enhances their performance, making them a promising solution for tackling pharmaceutical waste in wastewater treatment.

4.1.3 2D Nanomaterial

The discovery of graphene has led to the development of advanced 2D nanomaterials like graphene derivatives (GO, rGO), transition metal dichalcogenides (TMDs), and MXenes. These materials possess high surface areas, tunable layer thickness, and customizable surface chemistry, making them ideal for pharmaceutical waste removal from wastewater applications. 2D nanomaterials excel in adsorption, photocatalysis, and electrocatalysis due to their fast diffusion kinetics, abundant active sites, and stability in water. Functionalization and hybridization, such as with magnetic Fe_3O_4, enhance their dispersion stability and recyclability. In pharmaceutical waste removal, their high adsorption capacity allows efficient binding and removal of contaminants, while photocatalytic processes using ROS degrade pharmaceutical pollutants into harmless byproducts under light exposure. By optimizing properties like bandgap energy and charge mobility through doping or defect engineering, 2D nanomaterials provide eco-friendly, efficient, and scalable solutions for addressing pharmaceutical contamination in water. Recent advancements in 2D nanomaterials have demonstrated significant potential in adsorption applications, particularly for contaminants and antibiotics. For example, Zhang et al. [51] fabricated flower-like, wrinkled MoS_2 nanosheets via hydrothermal processing, achieving a specific surface area of 23.9 $m^2\ g^{-1}$ and a maximum adsorption capacity of 217 $mg\ g^{-1}$ for diclofenac, outperforming commercial MoS_2, activated carbon, and rGO. The adsorption kinetics followed a pseudo-second-order model, indicating chemical sorption as the underlying mechanism. Similarly, Goyal et al. [52] synthesized boron nitride (BN) nanosheets with a surface area of 900 $m^2\ g^{-1}$, achieving a removal efficiency of 95% for estrone within 2 h. The adsorption mechanism was also attributed to chemical sorption, as confirmed by XPS, with data fitting the Langmuir model. Further advancements in BN nanosheets by introducing defects and hierarchical pores significantly enhanced the adsorption capacity for tetracycline, as demonstrated by Wang et al. [53] with a capacity increase of 38% (up to 438 $mg\ g^{-1}$)—the introduction of boron vacancies manipulated electronic structures, creating more active sites and enhancing adsorption behavior. Moreover, few-layer g-C_3N_4 nanosheets exhibited adsorption and photocatalytic properties, with tetracycline removal efficiencies of 41% via adsorption and an additional 38% under illumination. The nanosheets' unique surface properties, including sp^2-hybridized carbon and delocalized bonds, contributed to their superior performance. In another study, Liu et al. [53] utilized a microwave-assisted hydrothermal approach to synthesize Cu-TCPP nanosheets, achieving adsorption capacities of 130 $mg\ g^{-1}$ for oxytocin and 150 $mg\ g^{-1}$ for tetracycline, reaching equilibrium conditions quickly. The adsorption data fitted Langmuir isotherms, implying monolayer adsorption. Li et al. [54] synthesized covalent triazine frameworks (CTFs) with specific surface areas of 565 $m^2\ g^{-1}$, which showed adsorption capacities of up to 351 $mg\ g^{-1}$ for nitrofurazone, following a pseudo-second-order model and Langmuir isotherms. Lastly, Jiang et al. [55] developed zwitterionic covalent organic frameworks (COFs) with ultrafast adsorption kinetics, removing 99.4% of norfloxacin in 30 s, achieving a maximum 824 $mg\ g^{-1}$ capacity.

The zwitterionic nature and nanochannel structures enhanced active site accessibility, enabling efficient monolayer adsorption. These studies collectively highlight the potential of tailored 2D nanomaterials in achieving high adsorption efficiencies through structural modifications and surface engineering.

4.1.4 3D Nanomaterials

Three-dimensional (3D) nanomaterials, characterized by their highly interconnected network structure that provides numerous active sites, have become increasingly popular in environmental remediation, including removing pharmaceutical contaminants from water. These materials, such as metal–organic frameworks (MOFs), mesoporous silica, and 3D carbon-based materials, are highly effective in photocatalytic and adsorption applications. Their large surface areas, tunable porosity, and diverse functional groups make them ideal candidates for efficiently removing pharmaceutical pollutants. In photocatalytic applications, 3D nanomaterials excel because they provide many active sites for adsorption and photocatalytic reactions. For example, MOFs, particularly those containing transition metals like copper (Cu), zinc (Zn), and iron (Fe), have been widely studied for their photocatalytic activity. In a study by Zhu et al. [56], a Cu-based MOF was utilized to remove antibiotics such as tetracycline and amoxicillin under UV light. The 3D framework of the MOF enabled efficient adsorption of pollutants, and its photocatalytic properties led to over 90% degradation of these contaminants within 60 min. Similarly, 3D porous carbon materials, such as graphene aerogels, have shown remarkable photocatalytic performance for the degradation of pharmaceutical contaminants. A study by Dhiman et al. [57] demonstrated that graphene aerogel-based composites achieved 85% removal of diclofenac under visible light, with enhanced photocatalytic activity due to the high surface area and efficient charge separation properties of the 3D network. In addition to photocatalysis, 3D nanomaterials are highly effective in adsorption processes. These materials' large surface area and tunable porosity provide numerous binding sites for pharmaceutical contaminants, enhancing their adsorption capacity. For instance, mesoporous silica materials have been extensively studied for their adsorption properties due to their well-defined structure and high surface area. In a study by Bui et al. [58], mesoporous silica was used to adsorb ibuprofen from water, achieving over 90% removal efficiency. Hydrogen bonding and van der Waals interactions between the surface of the mesoporous silica and the pharmaceutical molecules drove the adsorption. Similarly, 3D MOFs have been studied for their adsorption capabilities. A study by Gai et al. [59] demonstrated that a Zn-based MOF exhibited high adsorption efficiency for various antibiotics, including ciprofloxacin and levofloxacin, with removal efficiencies exceeding 80% in aqueous solutions. The unique structural features of 3D nanomaterials, such as their high surface area, tunable porosity, and ability to provide multiple active sites, make them highly effective for removing pharmaceutical contaminants. Their combined photocatalytic and adsorption properties enable them to degrade and adsorb a wide range of pollutants, including antibiotics, analgesics, and hormones, making them an ideal solution for wastewater treatment. Furthermore, integrating 3D nanomaterials with other materials can further enhance their performance, leading to even more efficient and sustainable solutions for environmental remediation.

4.1.5 Hybrid Nanomaterials

Hybrid nanomaterials, which combine two or more distinct materials (such as inorganic and organic compounds or two different nanomaterials) at the nanoscale, have emerged as promising materials for removing pharmaceutical contaminants. These materials take advantage of the complementary properties of the constituent components, offering enhanced photocatalytic and adsorption capabilities. The versatility of hybrid nanomaterials lies in their ability to combine the strengths of each material, such as the high surface area and tunable porosity of inorganic nanomaterials and the functional versatility of organic components. As a result, hybrid nanomaterials have been extensively explored for environmental remediation, particularly in the degradation and removal of pharmaceutical pollutants from water. In photocatalytic applications, hybrid nanomaterials have demonstrated significant improvements in efficiency due to their enhanced charge separation, light absorption, and reactive species generation. For example, hybrid nanocomposites formed by integrating carbon-based materials, such as carbon quantum dots (CQDs) or graphene oxide (GO), with semiconductors like TiO_2 or g-C_3N_4 have shown excellent photocatalytic performance for pharmaceutical degradation. A study by Wang et al. [60] demonstrated that a CQDs-TiO_2 hybrid photocatalyst achieved over 90% degradation of tetracycline under visible light irradiation. The CQDs extended the light absorption range of TiO_2, while the hybrid structure facilitated efficient charge transfer, leading to enhanced photocatalytic activity. Similarly, a graphene oxide (GO)-g-C_3N_4 hybrid system was used for ciprofloxacin degradation by Lakshmi CN et al. [61], achieving over 95% removal under visible light due to the synergistic effect of the GO and g-C_3N_4, which promoted the generation of reactive oxygen species (ROS) and improved the overall photocatalytic efficiency. In addition to photocatalysis, hybrid nanomaterials exhibit excellent adsorption properties, which are particularly useful in removing pharmaceutical pollutants from water. These materials combine the large surface area and high adsorption capacity of inorganic components like metal–organic frameworks (MOFs), mesoporous silica, and activated carbon with the functional groups of organic components such as CQDs, polymers, or peptides. For instance, a hybrid nanocomposite of N-doped CQDs and MOFs has achieved high adsorption efficiency for antibiotics like amoxicillin and tetracycline. In a study by Kumari et al. [62], the N-doped CQDs-MOF composite exhibited excellent adsorption performance, with over 90% removal of tetracycline within 60 min. The N-doped CQDs enhanced the adsorption by providing functional groups for interaction with the pollutants. They facilitated the regeneration of the MOF structure, making it a highly efficient material for wastewater treatment. Another example is the hybridization of magnetic nanoparticles with carbon-based materials for the adsorption of pharmaceuticals. Gu et al. [63] synthesized a magnetic carbon-based hybrid nanomaterial that effectively removed tetracycline from water. The hybrid material demonstrated a high adsorption capacity, and the magnetic properties allowed for easy recovery and reuse, making it an attractive option for practical applications in wastewater treatment. The synergistic effects of the carbon-based component and magnetic nanoparticles provided a dual-functionality system that combined high adsorption efficiency with ease of separation and regeneration. Integrating multiple materials into hybrid nanomaterials enables the development of multifunctional systems that not only degrade pharmaceutical contaminants but also remove them from water through efficient adsorption. The ability to tune the properties of hybrid nanomaterials, such as surface chemistry, porosity, and functionalization, makes them a versatile and promising solution for addressing pharmaceutical pollution in water. These materials hold great potential for large-scale applications in environmental remediation, offering sustainable and efficient methods for removing a wide range of pharmaceutical pollutants.

5 Recent Examples of Targeted Drug Removal from Wastewater by Nanomaterial

Table 1 represents some more advancement in recent years for particular drug removal from different types of nanomaterials.

6 Conclusion

With an emphasis on the importance of dimensionality, green synthesis, and hybrid materials, this chapter provided a thorough analysis of the synthesis, characteristics, mechanism, and uses of nanomaterials for the removal of pharmaceutical pollutants from wastewater. Future research in this area should focus on the commercialization and scalability of nanomaterial-based technologies as well as the development of nanomaterials with improved degradation capabilities. It should also look into the toxicity and environmental impact of these materials in conjunction with the current wastewater treatment infrastructure. More funds should be allocated to nanotechnology research and development. Periodically, a proper drug disposal campaign should be organized. To fully utilize nanotechnology and guarantee a better, healthier environment for coming generations, further research and innovation are essential.

Table 1 Examples of recent nanomaterial-based targeted pharmaceutical drug removal from wastewater

Drugs	Associated disease	Efficiency (%)	Removal method	References
Sulfadiazine	Antibiotics	94	Adsorption (Graphene-based adsorbent)	[64]
Levonorgestrel	Hormonal contraceptives	93	Adsorption (Magnetic Fe_3O_4 hybrid nanomaterial)	[65]
Progesterone	Hormone therapy	92	Adsorption (Mesoporous silica)	[66]
Paracetamol	Pain relievers (Analgesics)	90	Photocatalysis (g-C_3N_4 doped with metal oxides)	[67]
Atenolol	Beta-blockers (Cardiac therapy)	89	Photocatalysis (Graphene oxide hybrid)	[68]
Ketoprofen	Anti-inflammatory drugs (NSAIDs)	88	Adsorption (Mesoporous carbon material)	[69]
Acetaminophen	Pain relievers (Analgesics)	87	Photocatalysis (TiO_2–g-C_3N_4 nanocomposite)	[70]
Naproxen	Anti-inflammatory drugs (NSAIDs)	82	Adsorption (Functionalized CNTs)	[71]

References

1. Oluwole AO, Omotola EO, Olatunji OS (2020) Pharmaceuticals and personal care products in water and wastewater: a review of treatment processes and use of photocatalyst immobilized on functionalized carbon in AOP degradation. BMC Chem 14(1):62. https://doi.org/10.1186/s13065-020-00714-1
2. Fakhri BMS, Ghassemi Barghi N, Moradnia Mehdikhanmahaleh M, Raeis Zadeh SMM, Mousavi T, Rezaee R et al (2024) Pharmaceutical wastewater toxicity: an ignored threat to the public health. Sustain Environ 10(1). https://doi.org/10.1080/27658511.2024.2322821
3. Akintola AT, Ayankunle AY (2023) Improving pharmaceuticals removal at wastewater treatment plants using biochar: a review. Waste Biomass Valorization. https://doi.org/10.1007/s12649-023-02070-2
4. Estrada-Almeida AG, Castrejón-Godínez ML, Mussali-Galante P, Tovar-Sánchez E, Rodríguez A (2024) Pharmaceutical pollutants: ecotoxicological impacts and the use of agro-industrial waste for their removal from aquatic environments. J Xenobiotics 14(4):1465–518. https://www.mdpi.com/2039-4713/14/4/82
5. Ortúzar M, Esterhuizen M, Olicón-Hernández DR, González-López J, Aranda E (2022) Pharmaceutical pollution in aquatic environments: a concise review of environmental impacts and bioremediation systems. Front Microbiol 13. https://doi.org/10.3389/fmicb.2022.869332/full
6. Gkika DA, Vordos N, Mitropoulos AC, Lambropoulou DA, Kyzas GZ (2023) Nanomaterials and their properties: thermal analysis, physical, mechanical and chemical properties. In: Advanced and innovative approaches of environmental biotechnology in industrial wastewater

treatment. Singapore: Springer Nature Singapore. pp 301–31. https://doi.org/10.1007/978-981-99-2598-8_14
7. Guari Y (2024) Advanced porous nanomaterials: synthesis, properties, and applications. Nanomaterials 14(19):1602. https://www.mdpi.com/2079-4991/14/19/1602
8. Parashar M, Shukla VK, Singh R (2020) Metal oxides nanoparticles via sol-gel method: a review on synthesis, characterization and applications. J Mater Sci Mater Electron 31(5):3729–49. https://doi.org/10.1007/s10854-020-02994-8
9. Pandey S, Srivastava A, Rawat P, Chauhan SK, Ram A, Diwedi VK et al (2024) Eco-friendly synthesis of ZnO for efficient photodegradation of pharmaceutical drug removal by photocatalysis. ACS Omega. https://doi.org/10.1021/acsomega.4c06272
10. Algarra M, Jodeh S, Aqel I, Hanbali G, Radi S, Tighadouini S et al (2022) Phenylamine/amide grafted in silica as sensing nanocomposites for the removal of carbamazepine: A DFT approach. Chemosensors 10(2):76. https://www.mdpi.com/2227-9040/10/2/76
11. Pakowski Z (2007) Modern methods of drying nanomaterials. Transp Porous Media 66(1–2):19–27. https://doi.org/10.1007/s11242-006-9019-x
12. Wong TW, John P (2016) Advances in spray drying technology for nanoparticle formation. In: Handbook of nanoparticles. Cham: Springer International Publishing, pp 329–46. https://doi.org/10.1007/978-3-319-15338-4_18
13. Panneri S, Ganguly P, Mohan M, Nair BN, Mohamed AAP, Warrier KG et al (2017) Photoregenerable, bifunctional granules of carbon-doped g-C 3 N 4 as adsorptive photocatalyst for the efficient removal of tetracycline antibiotic. ACS Sustain Chem Eng 5(2):1610–8. https://doi.org/10.1021/acssuschemeng.6b02383
14. Mohammadi M, Hadadzadeh H, Kaikhosravi M, Farrokhpour H, Shakeri J (2020) Selective photocatalytic oxidation of benzyl alcohol at ambient conditions using spray-dried g-C3N4/TiO_2 granules. Mol Catal 490:110927. https://linkinghub.elsevier.com/retrieve/pii/S2468823120301826
15. Manawi YM, Ihsanullah, Samara A, Al-Ansari T, Atieh MA (2018) A review of carbon nanomaterials' synthesis via the chemical vapor deposition (CVD) method. Materials (Basel) 11(5):822. https://www.mdpi.com/1996-1944/11/5/822
16. Baig N, Kammakakam I, Falath W (2021) Nanomaterials: a review of synthesis methods, properties, recent progress, and challenges. Mater Adv 2(6):1821–71. https://xlink.rsc.org/?DOI=D0MA00807A
17. González-Poggini S, Rosenkranz A, Colet-Lagrille M (2021) Two-dimensional nanomaterials for the removal of pharmaceuticals from wastewater: a critical review. Processes 9(12):2160. https://www.mdpi.com/2227-9717/9/12/2160
18. Demazeau G (2011) Solvothermal and hydrothermal processes: the main physico-chemical factors involved and new trends. Res Chem Intermed 37(2–5):107–23. https://doi.org/10.1007/s11164-011-0240-z
19. Yoshimura M, Byrappa K (2008) Hydrothermal processing of materials: past, present and future. J Mater Sci 43(7):2085–103. https://doi.org/10.1007/s10853-007-1853-x
20. Malakootian M, Nasiri A, Asadipour A, Kargar E (2019) Facile and green synthesis of ZnFe2O4@CMC as a new magnetic nanophotocatalyst for ciprofloxacin degradation from aqueous media. Process Saf Environ Prot 129:138–51. https://linkinghub.elsevier.com/retrieve/pii/S0957582019305750
21. Mugunthan E, Saidutta MB, Jagadeeshbabu PE (2019) Photocatalytic activity of ZnO-WO3 for diclofenac degradation under visible light irradiation. J Photochem Photobiol A Chem 383:111993. https://linkinghub.elsevier.com/retrieve/pii/S1010603019300814
22. Wei R, Wang P, Zhang G, Wang N, Zheng T (2020) Microwave-responsive catalysts for wastewater treatment: a review. Chem Eng J 382:122781. https://linkinghub.elsevier.com/retrieve/pii/S1385894719321916
23. Hasanpoor M, Aliofkhazraei M, Delavari H (2015) Microwave-assisted synthesis of zinc oxide nanoparticles. Procedia Mater Sci 11:320–5. https://linkinghub.elsevier.com/retrieve/pii/S2211812815004435

24. Gandamalla A, Manchala S, Verma A, Fu YP, Shanker V (2021) Microwave-assisted synthesis of ZnAl-LDH/g-C_3N_4 composite for degradation of antibiotic ciprofloxacin under visible-light illumination. Chemosphere 283:131182. https://linkinghub.elsevier.com/retrieve/pii/S0045653521016544
25. Tamaddon F, Mosslemin MH, Asadipour A, Gharaghani MA, Nasiri A (2020) Microwave-assisted preparation of ZnFe2O4@methyl cellulose as a new nano-biomagnetic photocatalyst for photodegradation of metronidazole. Int J Biol Macromol 154:1036–49. https://linkinghub.elsevier.com/retrieve/pii/S0141813019371296
26. Hujjatul Islam M, Paul MTY, Burheim OS, Pollet BG (2019) Recent developments in the sonoelectrochemical synthesis of nanomaterials. Ultrason Sonochem 59:104711. https://linkinghub.elsevier.com/retrieve/pii/S1350417719305929
27. Jafari K, Heidari M, Rahmanian O (2018) Wastewater treatment for Amoxicillin removal using magnetic adsorbent synthesized by ultrasound process. Ultrason Sonochem 45:248–56. https://linkinghub.elsevier.com/retrieve/pii/S1350417718301706
28. Dinesh GK, Pramod M, Chakma S (2020) Sonochemical synthesis of amphoteric Cu0-Nanoparticles using Hibiscus rosa-sinensis extract and their applications for degradation of 5-fluorouracil and lovastatin drugs. J Hazard Mater 399:123035. https://linkinghub.elsevier.com/retrieve/pii/S0304389420310244
29. Lv Y, Ma J, Liu K, Jiang Y, Yang G, Liu Y et al (2021) Rapid elimination of trace bisphenol pollutants with porous β-cyclodextrin modified cellulose nanofibrous membrane in water: adsorption behavior and mechanism. J Hazard Mater 403:123666. https://linkinghub.elsevier.com/retrieve/pii/S0304389420316526
30. Kebede TG, Dube S, Nindi MM (2019) Biopolymer electrospun nanofibres for the adsorption of pharmaceuticals from water systems. J Environ Chem Eng 7(5):103330. https://linkinghub.elsevier.com/retrieve/pii/S2213343719304531
31. Chauhan A, Sillu D, Agnihotri S (2019) Removal of pharmaceutical contaminants in wastewater using nanomaterials: a comprehensive review. Curr Drug Metab 20(6):483–505. http://www.eurekaselect.com/167763/article
32. Cai Z, Dwivedi AD, Lee WN, Zhao X, Liu W, Sillanpää M et al (2018) Application of nanotechnologies for removing pharmaceutically active compounds from water: development and future trends. Environ Sci Nano 5(1):27–47. https://xlink.rsc.org/?DOI=C7EN00644F
33. Singh S, Kumar V, Anil AG, Kapoor D, Khasnabis S, Shekar S et al (2021) Adsorption and detoxification of pharmaceutical compounds from wastewater using nanomaterials: a review on mechanism, kinetics, valorization and circular economy. J Environ Manage 300:113569. https://linkinghub.elsevier.com/retrieve/pii/S0301479721016315
34. Husein DZ, Hassanien R, Al-Hakkani MF (2019) Green-synthesized copper nano-adsorbent for the removal of pharmaceutical pollutants from real wastewater samples. Heliyon 5(8):e02339. https://linkinghub.elsevier.com/retrieve/pii/S2405844019359997
35. El-Maraghy CM, El-Borady OM, El-Naem OA (2020) Effective removal of levofloxacin from pharmaceutical wastewater using synthesized zinc oxid, graphen oxid nanoparticles compared with their combination. Sci Rep 10(1):5914. https://www.nature.com/articles/s41598-020-61742-4
36. Bagheri S, TermehYousefi A (2017) Do TO. Photocatalytic pathway toward degradation of environmental pharmaceutical pollutants: structure, kinetics and mechanism approach. Catal Sci Technol 7(20):4548–69. https://xlink.rsc.org/?DOI=C7CY00468K
37. Kaur T, Sraw A, Toor AP, Wanchoo RK (2016) Utilization of solar energy for the degradation of carbendazim and propiconazole by Fe doped TiO_2. Sol Energy 125:65–76. https://linkinghub.elsevier.com/retrieve/pii/S0038092X1500660X
38. El-Fawal EM, Younis SA, Zaki T. Designing $AgFeO_2$-graphene/Cu_2(BTC)3 MOF heterojunction photocatalysts for enhanced treatment of pharmaceutical wastewater under sunlight. J Photochem Photobiol A Chem 401:112746. https://linkinghub.elsevier.com/retrieve/pii/S1010603020305451
39. Rasheed T, Rizwan K, Shafi S, Bilal M (2022) Nanobiodegradation of pharmaceutical pollutants. In: Biodegradation and biodeterioration at the nanoscale. Elsevier, pp 635–53. https://linkinghub.elsevier.com/retrieve/pii/B9780128239704000269

40. Ji C, Nguyen LN, Hou J, Hai FI, Chen V (2017) Direct immobilization of laccase on titania nanoparticles from crude enzyme extracts of P. Ostreatus culture for micro-pollutant degradation. Sep Purif Technol 178:215–23. https://linkinghub.elsevier.com/retrieve/pii/S1383586616325047
41. Xu R, Tang R, Zhou Q, Li F, Zhang B (2015) Enhancement of catalytic activity of immobilized laccase for diclofenac biodegradation by carbon nanotubes. Chem Eng J 262:88–95. https://linkinghub.elsevier.com/retrieve/pii/S1385894714012686
42. Sendão RMS, Esteves da Silva JCG, Pinto da Silva L (2022) Photocatalytic removal of pharmaceutical water pollutants by TiO_2–Carbon dots nanocomposites: a review. Chemosphere 301:134731. https://linkinghub.elsevier.com/retrieve/pii/S0045653522012243
43. Yang Z, Yan J, Lian J, Xu H, She X, Li H. g-C_3N_4/TiO_2 nanocomposites for degradation of ciprofloxacin under visible light irradiation. Chem Select 1(18):5679–85. https://doi.org/10.1002/slct.201600861
44. Zhang J, Li Z, Lei Q, Zhong D, Ke Y, Liu W et al (2023) Significantly activated persulfate by novel carbon quantum dots-modified N-BiOCl for complete degradation of bisphenol-A under visible light irradiation. Sci Total Environ 870:161804. https://linkinghub.elsevier.com/retrieve/pii/S0048969723004199
45. Lin X, Song MH, Li W, Wei W, Wu X, Mao J et al (2024) Optimized design of quaternary amino-functionalized chitosan fibers for ultra-high diclofenac adsorption from wastewater. Chemosphere 357:141970. https://linkinghub.elsevier.com/retrieve/pii/S0045653524008634
46. Cao J, Dong X, Xie C, Xiao C, Wu Y, Shoulian W (2024) Preparation of nitrogen-doped magnetic carbon microspheres and their adsorption and degradation properties of tetracycline hydrochloride. Chem Eng Sci 300:120564. https://linkinghub.elsevier.com/retrieve/pii/S0009250924008649
47. Guo J, Gan W, Ding C, Lu Y, Li J, Qi S et al (2022) Black phosphorus quantum dots and Ag nanoparticles co-modified TiO_2 nanorod arrays as powerful photocatalyst for tetracycline hydrochloride degradation: pathways, toxicity assessment, and mechanism insight. Sep Purif Technol 297:121454. https://linkinghub.elsevier.com/retrieve/pii/S1383586622010115
48. Hassani A, Khataee A, Karaca S, Karaca C, Gholami P (2017) Sonocatalytic degradation of ciprofloxacin using synthesized TiO_2 nanoparticles on montmorillonite. Ultrason Sonochem 35:251–62. https://linkinghub.elsevier.com/retrieve/pii/S135041771630339X
49. Valadez-Renteria E, Perez-Gonzalez R, Gomez-Solis C, Diaz-Torres LA, Encinas A, Oliva J et al (2023) A novel and stretchable carbon-nanotube/Ni@TiO_2:W photocatalytic composite for the complete removal of diclofenac drug from the drinking water. J Environ Sci 126:575–89. https://linkinghub.elsevier.com/retrieve/pii/S1001074222002820
50. Oyetade OA, Martincigh BS, Skelton AA (2018) Interplay between electrostatic and hydrophobic interactions in the pH-dependent adsorption of ibuprofen onto acid-functionalized multiwalled carbon nanotubes. J Phys Chem C 122(39):22556–68. https://doi.org/10.1021/acs.jpcc.8b06841
51. Zhang Y, Yin Z, Dai C, Zhou X, Chen W (2016) Interfacial thermodynamics and kinetics of sorption of diclofenac on prepared high performance flower-like MoS_2. J Colloid Interface Sci 481:210–9. https://linkinghub.elsevier.com/retrieve/pii/S0021979716305057
52. Goyal A, Aggarwal D, Kapoor S, Goel N, Singhal S, Shukla J (2020) A comprehensive experimental and theoretical study on BN nanosheets for the adsorption of pharmaceutical drugs. New J Chem 44(10):3985–97. https://xlink.rsc.org/?DOI=C9NJ06029D
53. Wang S, Jia F, Kumar P, Zhou A, Hu L, Shao X, et al (2020) Hierarchical porous boron nitride nanosheets with versatile adsorption for water treatment. Colloids Surfaces A Physicochem Eng Asp 598:124865. https://linkinghub.elsevier.com/retrieve/pii/S0927775720304581
54. Li J, Liu A, Ji M, Chen T, Ma D (2024) Fluorescent detection of nitrofuran antibiotics with a recordly brilliant covalent organic framework. Microchem J 204:111049. https://linkinghub.elsevier.com/retrieve/pii/S0026265X24011615

55. Jiang W, Cui WR, Liang RP, Qiu JD (2021) Zwitterionic surface charge regulation in ionic covalent organic nanosheets: synergistic adsorption of fluoroquinolone antibiotics. Chem Eng J 417:128034. https://linkinghub.elsevier.com/retrieve/pii/S1385894720341528
56. Zhu S, Sun C, Zhu Z, Qu J, Fang Z, Chen Y et al (2024) Copper-based photocatalysts with natural organic ligands for efficient removal of tetracycline under visible light. J Environ Manage 370:123052. https://linkinghub.elsevier.com/retrieve/pii/S030147972403038X
57. Dhiman N, Tripathi VK, Dwivedi J, Gupta RK, Tripathi KM (2024) Photoactive graphene aerogel from biomass for the visible-light-induced degradation of pharmaceutical residues. ACS Sustain Resour Manag 1(6):1068–75. https://doi.org/10.1021/acssusresmgt.3c00088
58. Bui TX, Choi (2009) H Adsorptive removal of selected pharmaceuticals by mesoporous silica SBA-15. J Hazard Mater 168(2–3):602–8. https://linkinghub.elsevier.com/retrieve/pii/S0304389409002647
59. Gai S, Zhang J, Fan R, Xing K, Chen W, Zhu K et al (2020) Highly stable zinc-based metal–organic frameworks and corresponding flexible composites for removal and detection of antibiotics in water. ACS Appl Mater Interfaces 12(7):8650–62. https://doi.org/10.1021/acsami.9b19583
60. Wang Y, Chen H, Jia Z, Lv J, Lv Y, Wu J (2023) Analysis of photocatalytic properties of NS-CQDs/g-C_3N_4 composites. Micromachines 14(12):2143. https://www.mdpi.com/2072-666X/14/12/2143
61. Naga Lakshmi C, Irfan M, Sinha R, Singh N (2024) Magnetically recoverable Ni-doped iron oxide/graphitic carbon nitride nanocomposites for the improved photocatalytic degradation of ciprofloxacin: investigation of degradation pathways. Environ Res 242:117812. https://linkinghub.elsevier.com/retrieve/pii/S0013935123026166
62. Kumari K, Kadian P, Kumar M, Randhawa JK (2024) Understanding Norfloxacin's turn-on detection mechanism: focusing on amorphous carbon-coated iron oxide nanoparticles. ACS Appl Opt Mater 2(7):1444–53. https://doi.org/10.1021/acsaom.4c00206
63. Gu W, Huang X, Tian Y, Cao M, Zhou L, Zhou Y et al (2021) High-efficiency adsorption of tetracycline by cooperation of carbon and iron in a magnetic Fe/porous carbon hybrid with effective Fenton regeneration. Appl Surf Sci 538:147813. https://linkinghub.elsevier.com/retrieve/pii/S0169433220325708
64. Zhang L, Wang Y, Jin S, Lu Q, Ji J (2017) Adsorption isotherm, kinetic and mechanism of expanded graphite for sulfadiazine antibiotics removal from aqueous solutions. Environ Technol 38(20):2629–38. https://doi.org/10.1080/09593330.2016.1272637
65. Aguinaga Martínez MV, Llamas NE, Ávila Orozco FD, Domini CE, Acebal CC (2020) Oxidized carbon nanotubes as sorbent for miniaturized solid-phase extraction of progestins from environmental water samples prior to their determination by HPLC-UV. Microchim Acta 187(2):153. https://doi.org/10.1007/s00604-020-4116-z
66. Carvalho LB, Chagas PMB, Marques TR, Razafitianamaharavo A, Pelletier M, Nolis P et al (2019) Removal of the synthetic hormone methyltestosterone from aqueous solution using a β-cyclodextrin/silica composite. J Environ Chem Eng 7(6):103492. https://linkinghub.elsevier.com/retrieve/pii/S2213343719306153
67. Puri S, Thakur I, Verma A, Barman S (2021) Degradation of pharmaceutical drug paracetamol via UV irradiation using Fe-TiO_2 composite photocatalyst: statistical analysis and parametric optimization. Environ Sci Pollut Res 28(34):47327–41. https://doi.org/10.1007/s11356-021-13895-6
68. Kyzas GZ, Koltsakidou A, Nanaki SG, Bikiaris DN, Lambropoulou DA (2015) Removal of beta-blockers from aqueous media by adsorption onto graphene oxide. Sci Total Environ 537:411–20. https://linkinghub.elsevier.com/retrieve/pii/S004896971530485X
69. Georgin J, Franco DSP, da Boit Martinello K, Lima EC, Silva LFO (2022) A review of the toxicology presence and removal of ketoprofen through adsorption technology. J Environ Chem Eng 10(3):107798. https://linkinghub.elsevier.com/retrieve/pii/S2213343722006716
70. Zyoud AH, Zubi A, Hejjawi S, Zyoud SH, Helal MH, Zyoud SH et al (2020) Removal of acetaminophen from water by simulated solar light photodegradation with ZnO and TiO_2 nanoparticles: catalytic efficiency assessment for future prospects. J Environ Chem Eng 8(4):104038. https://linkinghub.elsevier.com/retrieve/pii/S2213343720303869

71. Kurwadkar S, Hoang TV., Malwade K, Kanel SR, Harper WF, Struckhoff G (2019) Application of carbon nanotubes for removal of emerging contaminants of concern in engineered water and wastewater treatment systems. Nanotechnol Environ Eng 4(1):12. https://doi.org/10.1007/s41204-019-0059-1

Nanomaterials for the Removal of Dyes from Wastewater

Arshad Iqbal, Naseem Ahmad, Iqbal Azad, Mohsin Vahid Khan, Mohammad Shahadat, Mohd Arshad, and Nafees Ahmad

Abstract Dyes are frequently used to color objects in industries such as textiles, tanning, leather, cosmetics, and many more. These industries release huge amounts of wastewater that pose a risk to the environment and human well-being. Therefore, dye removal from contaminated water has received significant attention. For this purpose, adsorption and photocatalysis have been proven to be efficient and cost effective for removing dyes than the conventional biological, chemical, and physical processes. The utilization of nanomaterials through a combination of photocatalytic and adsorption has emerged to be an efficient method. The present chapter provides a comprehensive out-looks of dye removal treatment including their classification, toxicity, and dye elimination methods with especial focus on nanomaterials. Based on the reported date, it is established that nanomaterials have demonstrated significant dye removal potential to treat industrial wastewater.

Keywords Nanomaterials · Dye removal · Wastewater treatment · Adsorption · Photocatalysis

---

A. Iqbal
Department of Physics, Integral University, Lucknow, India

N. Ahmad · I. Azad · N. Ahmad (✉)
Department of Chemistry, Integral University, Lucknow, India
e-mail: siddiquenafees123@gmail.com

M. V. Khan
Department of Biosciences, Integral University, Lucknow, India

M. Shahadat
Department of Textile and Fiber Engineering, Indian Institute of Technology, New Delhi, India

Institute of Oceanography and Environment, Universiti Malaysia Terengganu, Kuala Nerus, Terengganu, Malaysia

M. Arshad
Saiyyid Hamid Senior Secondary School SHSSS (Boys), Aligarh Muslim University, Aligarh, Uttar Pradesh, India

S. A. Bhawani and M. Jawaid (eds.), *Nanomaterials for Separation of Hazardous Contaminants from Wastewater*, Smart Nanomaterials Technology,
https://doi.org/10.1007/978-981-95-6079-0_10

1 Introduction

One of the most valuable resources is water, which is an essential and vital component of life. Just 0.007% of all water is fit for human consumption, even though 71% of the surface of the earth is occupied by water, of which almost 97.5% is salted and only 2.5% is fresh [1]. The primary cause of water contamination is industrial waste, which contains a variety of dyes and other contaminants [2, 3]. Industries like textiles, leather tanning, cosmetics, etc., have made substantial use of dyes [4–6]. Synthetic dyes are extensively used in industries, for example leather, paper, and textiles, contributing significantly to water pollution. These dyes are poisonous, non-biodegradable in nature, persistent, posing severe risks to aquatic species and human health. The dyes are visible at very low concentrations and have a recalcitrant nature that can cause unintended harm to aquatic life, for example reducing the penetration of sunlight and resisting the reactions of photocatalysis, their presence in the hydrosphere poses a serious pollution risk. Their waste products formed through the dyeing processes are unsafe to the health of humans, mutagenic, hazardous, and increase the levels of biochemical oxygen demand (BOD) and chemical oxygen demand (COD) in aquatic sources [7–9]. Therefore, before wastewater is finally met into the environment, dyes must be effectively removed. Numerous approaches have been used for the removal of dyes like separation through membrane, progressive oxidation, aerobic or anaerobic action, electrochemical methods, and flocculation-coagulation [10–14]. However, these methods often fall short in terms of efficiency and sustainability. Several studies have been reported on the degradation of and removal of dyes via adsorption and photocatalysis. Hazardous inorganic and organic contaminants found in wastewater have been successfully reduced by the use of adsorption, one of the most efficient wastewater treatment techniques [15, 16]. There have been several reports on the viability of using inexpensive adsorbents to extract dyes from wastewater. Chitosan [17], wheat straw [18], orange and banana peels [19], oil palm ash [20], wasted tea leaves [21], degreased coffee bean [22], palm kernel shell [23], coconut shell [24], fly ash [25], and mango peels [26] are a few of these. However, the usage of the above mentioned adsorbents has been restricted due to the inconvenience of separation. Therefore more research is needed to utilize the promising adsorbents. The emergence of nanomaterials (NMs) offers a viable alternative providing innovative methods for the removal of dye due to their characteristics such as high surface area, tunable porosity, excellent adsorption, and catalytic capabilities. Nanomaterials, which include devices and structures with sizes ranging from 1 to 100 nm, are very significant [27, 28]. Numerous nanomaterials, including graphene or graphene-based nanomaterials, nanometals, nanometal oxides, and polymer-based nanoparticles, have been extensively applied to remove pollutants from wastewater [29–32]. The chapter deals with the description of dye kinds and toxicity, various approaches used in wastewater treatment and the use of nanomaterials for photocatalytic degradation and adsorption.

2 Classification of Dyes in Wastewater

In fundamental nature, dyes are organic substances that can adhere to cloth surfaces to provide color [8]. The ingredients, colors, and applications of dyes can all be used to classify them; the most popular approach is based on uses [11]. Few examples of types of dyes according to their chemical structure such as triarylmethane, indigoid, anthraquinone, azo, phthalocyanine, nitro, and nitroso dyes are presented in Table 1. The most significant categories of these dyes are triarylmethane, anthraquinone, and azo dyes.

Furthermore, based on the particle charges dissolve in aqueous solutions, the dyes utilized in the textile industry are classified as non-ionic (disperse dyes), anionic (direct, acid, and reactive dyes), and cationic (all basic dyes) [33]. Due to their intricate structure, dyes are useless when exposed to heat, light, bacteria, or even oxidizing chemicals which make dyes difficult to degrade. Most of the dyes prevent photosynthesis in aquatic plants and animals by reducing water quality and altering sunlight penetration and oxygen solubility that result in serious health problems for people, including tissue necrosis, quadriplegia, jaundice, cyanosis, Heinz body formation, vomiting, shock, and elevated heart rate [34, 35].

2.1 *Azo Dyes*

Monoazo, diazo, triazo, and polyazo dyes make up the bulk of artificial aromatic dyes. The naphthalene and benzene rings to which these dyes are linked are regularly replaced with other functional groups, including sulfonate, nitro, methyl, hydroxyl, chloro, and triazine amine. These dyes possess one, two, three, or more (N=N) groups [36]. Azo dyes are broadly utilized in paper printing, food coloring, cosmetics, pharmaceuticals, and textile dyeing. Because of their great durability, affordability, and range of hues, almost 80% of azo dyes are utilized in the textile industry [37]. According to estimates, 10–15% of colors in wastes are discharged into receiving

Table 1 Dye class identified by chemical structure [8]

Types of dyes	Applications of dyes	Toxicity	Example of dyes
Acid	Wool, silk, leather	Carcinogenic	Methyl orange, sunset yellow
Cationic	Paper, modified polyester	Carcinogenic	Methylene blue, rhodamine 6G
Reactive	Cotton, wool, nylon	Dermatitis, allergic	Reactive red 120 Reactive red 198
Disperse	Acrylic fiber	Allergenic	Disperse orange 3
Direct	Paper, cotton	Cancerous	Congo red, direct red 23

water bodies after becoming unbound with textile fibers [38]. Diflurochloropyrimidine, trichloropyrimidine, chlorotriazine, and vinyl sulfone are examples of responsive dyes, which are azo dyes that make covalent connections among their reactive groups [39]. Reactive dyes are superior to other dye classes in the textile industry due to their low consumption of energy, simplicity of application, water-fastness, and vivid color. Reactive dyes are mutagenic and carcinogenic in nature and their mineralization leads to an arylamine and aromatic amine have serious negative effects on human health but still reactive dye are widely used in the industries.

2.2 *Anthraquinone Dyes*

The textile industry has made extensive use of anthraquinone dyes due to their broad variety of color hues, fastness characteristics, ease of application, and consumption of low energy [40]. They are the largest class of textile dyes after azo dyes. The high affinity of anthraquinone dyes for silk and wool is another significant benefit of using them in dyeing procedures. However, anthraquinone dyes have a reduced rate of decolorization because of their fused aromatic structure. The removal of anthraquinone dyes has taken great interest due to most of these dyes are carcinogenic, mutagenic, and poisonous.

2.3 *Triarylmethane Dyes*

Triphenylmethane dyes are aromatic xenobiotics made composed of one p-quinoid group (chromophore) and two benzene rings joined to a central carbon atom. $-NH_2$, $-NR_2$, and –OH are the most frequent auxochromes. These dyes, which include bromophenol blue (BB), malachite green (MG), crystal violet (CV), and others, exhibit vivid hues with classic red, violet, blue, and green tones [41]. In the textile industry, these dyes are frequently used to color substrates such as cotton, wool, silk, polyacrylonitrile, and modified nylon. In addition, triphenylmethane dyes are used in leather, gasoline, and paper [42]. Their strong affinity, poor light fastness, and very inexpensive dyeing especially when washed slowly and exposed to light make them appropriate for the coloring of textile substrates. Additionally, triphenylmethane dyes have antiprotozoal, antifungal, and antibacterial properties; they are frequently utilized in pharmaceuticals as therapeutic disinfectants. These colors can be used to treat diabetes and clean surgical incisions [43]. Some of the example of these dyes, such fuchsine, phenolphthalein, and fluorescein, can be used as indicator dyes because of their sensitivity to pH.

3 Properties of Nanomaterials for Dye Removal

Nanomaterials due to their nanoscale size have unique physicochemical properties that make them perfect for extracting dye from aqueous solutions. One of their most crucial features is their high surface area-to-volume ratio, which provides dye molecules with more active sites to bind to and boosts adsorption efficacy [44]. Additionally, chemically functionalizing nanoparticles can achieve the selective adsorption of specific colors. Customizing nanomaterials to target certain dye molecules by changing their surface chemistry can improve the removal process' efficacy [45]. The properties of nanomaterials to catalyze the breakdown of complicated dye compounds into simpler, non-toxic byproducts is another essential characteristic.

Furthermore, number of nanomaterials with photocatalytic qualities, which allow them to employ light energy to speed up the degradation of dyes. When exposed to light, the nanomaterials produce reactive oxygen species (ROS), which oxidize and degrade dye molecules, making them an environmentally benign and energy-efficient dye removal method [46, 47]. Photocatalytic technique is particularly beneficial for cleaning industrial effluents by mineralizing dyes that contain organic colors. All these characteristics show the removal and detoxification of dyes in wastewater by improving adsorption efficiency and facilitating the breakdown of dyes into innocuous byproducts. Nanomaterials are therefore being investigated more and more as cutting-edge components for effective and environmentally friendly water filtration systems.

4 Types of Nanomaterials Used for Dye Removal

Nanomaterials due to their special qualities, such as their large surface area, functionality, and effective adsorption, have become efficient photocatalysts for the dye removal. There are several nanomaterials, such as metal oxides, carbon-based nanomaterials, nanocomposites, metal organic frameworks (MOFs) and biopolymers, emphasizing their benefits and drawbacks in tackling dye removal. Below is the table summarizing the advantage and disadvantage of various nanomaterials for the removal of dye (Table 2).

4.1 Carbon-Based Nanomaterials

The excellent characteristics of carbon-based nanostructures make them suitable catalysts for dye removal. The high adsorption capacity of activated carbon nanoparticles is a result of their extremely porous structure, which gives dye molecules plenty of surface area to bind with the surface of carbon-based material. Graphene oxide (GO) is a carbon rich materials, having functional groups like hydroxyl, epoxy, and

Table 2 Examples of few nanomaterials with their advantages and disadvantages in the removal of dyes

Nanomaterial	Pros	Cons
Carbon nanotubes (CNTs)	• High surface area and adsorption capacity • Excellent mechanical strength • Can be functionalized for enhanced performance	• Expensive to produce • Potential environmental toxicity • Difficult to separate after use
Graphene oxide (GO)	• High adsorption capacity for dyes • Easy to functionalize • Good dispersibility in water	• Complex synthesis process • Potential environmental concerns • High cost
Metal oxides (e.g., TiO_2, ZnO)	• Photocatalytic properties for dye degradation • High stability and reusability • Cost effective	• Limited adsorption capacity • Potential toxicity of nanoparticles • Requires UV light for activation (for TiO_2)
Magnetic nanoparticles (e.g., Fe_3O_4)	• Easy separation using magnetic fields • High surface area • Can be functionalized for specific dyes	• May require surface modification • Potential aggregation issues • Limited reusability
Nanoclays	• Low cost and abundant • High adsorption capacity • Environmentally friendly	• Limited selectivity for specific dyes • May require modification for better performance • Slower adsorption kinetics
Dendrimers	• High selectivity for specific dyes • Tunable surface chemistry • High adsorption capacity	• Expensive to synthesize • Complex synthesis process • Limited scalability
Chitosan nanoparticles	• Biodegradable and eco-friendly • High adsorption capacity • Low cost	• Limited stability in acidic conditions • Requires cross-linking for improved performance • Moderate reusability
Metal-organic frameworks (MOFs)	• Extremely high surface area • Tunable pore size and functionality • High adsorption capacity	• Expensive to produce • Limited stability in water • Complex synthesis process

carboxyl on its surface that allow for strong interactions with dye molecules via π–π stacking, hydrogen bonds, and electrostatic forces [48, 49]. Additionally, carbon nanotubes (CNTs) are very effective in adsorbing dyes because of their remarkable mechanical strength and surface area [50, 51]. These carbon-based materials when used for the adsorption of dyes shows effective results in wastewater treatment.

4.2 Metal Oxide Nanoparticles

Metal oxide nanoparticles due to their special characteristics are very effective in removing dyes. A common photocatalyst, titanium dioxide (TiO_2) effectively breaks down dye molecules when exposed to ultraviolet light, producing reactive oxygen species that destroy contaminants. Zinc oxide (ZnO) can be used to treat contaminated water since it has antibacterial qualities in addition to photocatalytic activity when exposed to UV light. Post-treatment procedures are made simpler by iron oxide's (Fe_3O_4) unique magnetic properties, which allow for simple separation from treated water using an external magnetic field [52–54]. Apart from these nanomaterials, other metal oxide nanomaterials such as NiO, Sb_2S_3, CuO, etc., are used for the treatment of wastewater treatment applications by degrading dyes.

4.3 Polymeric Nanomaterials

The specific features of polymer-based nanoparticles make them very effective for the removal of a wide range of dyes. Due to their high affinity for dye molecules, nanostructured polymers provide effective adsorption via hydrophobic effects, hydrogen bonds, and electrostatic interactions [55–57]. To improve these polymeric materials' ability to bind dyes, it can be doped with nanoparticles or particular functional groups can be incorporated. In order to enable selective and effective removal, the polymers are specifically engineered with binding sites that correspond to the size, shape, and functional groups of the target dye molecules. Because of its specificity, polymeric nanomaterials are suitable catalysts for the complex wastewater that contains several dyes, providing a flexible and efficient solution for dye removal applications.

5 Mechanisms of Dye Removal Using Nanomaterials

5.1 Adsorption

Adsorption is the process by which the boundary between the liquid (micropollutant molecule) and photocatalysts are concentrated on the region of a porous substance. The substance that comes into contact with a solid is called an adsorbent, while the substance that accumulates at the interface is called an adsorbate [58]. Depending on the mechanism of adsorbate species interaction with the adsorbent surface, the adsorption process can be roughly divided into two categories, i.e., chemical sorption and physical sorption. During physical adsorption, the adsorbate sticks to the adsorbents by means of weak physical forces like π–π interactions, dipole interactions, static interactions, polarity, hydrogen bonding, hydrophobicity, and van der Waals forces [59]. Adsorbates are chemically bound to the surface of an adsorbent in chemical adsorption by the force of electron exchange. Since it has been demonstrated to be most efficient method for eliminating colors from wastewater when cost, efficiency, and design flexibility are taken into account, the adsorption method is a widely used alternative for environmental cleanup. Additionally, adsorption does not create any hazardous elements and avoids subsequent contamination. An affordable and environmentally acceptable adsorbent is selected for the adsorption process based on the adsorption capacity, the likely cost of material reuse, and the surface area [60]. The porous catalysts with a large surface area of 500–2000 m^2/g and excellent adsorption proficiencies for a wide range of pollutants. These extraordinary characteristics are the result of the source of natural resources. However, it is crucial to search for less expensive and renewable precursors because coal and wood are fairly expensive natural resources. Agricultural waste and byproducts, such as chitosan, oil palm ash, banana and orange peels, leftover tea leaves, and degreased coffee beans, are the main sources of activated carbon used for the adsorption purpose. A study by Tan et al. suggests that MB may be eliminated by activated carbon derived from coconut husk [61]. The adsorption capacity decreased from 434.78 to 384.62 mg/g when the temperature was increased from 30 to 50 °C, suggesting that the adsorption reaction is temperature dependant. The greatest quantity of MB that can be adsorbed onto an adsorbent with a relatively large surface area of 1940 m^2/g and a full pore volume of 1.143 cm^3/g is 434.78 mg/g, based on the Langmuir isotherm calculations. Another study was carried out by Hameed et al. [62] on the adsorption capability of activated carbon derived from rattan sawdust. The results were consistent with pseudo-second-order kinetics, and the equilibrium was attained in less than 24 h. According to the Langmuir model, the equilibrium data showed that dye molecules were covered in a monolayer at the outer surface of the sawdust carbon. Apart from the naturally derived material used for the adsorption purpose, nanomaterial are also effective in the removal of micropollutant from wastewater. Nanomaterials, including semiconductor metal oxides, perovskites materials, etc., are used for the removal of wide range of dyes and other pollutants. Nanomaterials such as magnetic Fe_3O_4 [63], MgO [64], titanium dioxide (TiO_2) [65], zinc oxide (ZnO) [66], nanozerovalent iron [67], and nanozerovalent zinc [68], offer specific affinity and high surface area. These nanomaterials have minimal environmental impact, weak solubility, and no secondary contamination. Because of their superiority, they have also been selected to eradicate micropollutant. Figure 1 shows the adsorption of dyes on the porous surface of the nanoparticles.

5.2 *Photocatalysis*

Light energy is used by nanomaterials such as ZnO and TiO_2 to produce reactive oxygen species (ROS), which break down dye molecules into non-toxic substances. The use of advanced oxidation processes (AOPs) for the effective removal of a wide variety of organic dyes has lately attracted special attention due to the production of highly reactive hydroxyl radicals. AOPs include reactions such as ultrasonication,

Fig. 1 Adsorption of dyes on the surface of nanoparticles

photocatalysis, UV/H_2O_2, Fenton and photo-Fenton processes (Fe^{2+}/H_2O_2 and Fe^{2+}/H_2O_2/UV), and electrochemical activity [69]. The most efficient and least energy-intensive of these is photocatalytic degradation. For the photocatalysis, various nanomaterials such as titanates, perovskites, zinc oxide, and titanium dioxide are used for the degradation of dyes. The charge carriers produced by photogeneration, which comprise electrons (e^-) and holes (h^+), lead to the formation of ROS and subsequently the degradation of dyes. Electrons on the surface of photocatalysts are driven to the conduction band (CB) and holes are created in the valance band (VB) when the photocatalyst is subjected to radiation [70, 71]. While the hole oxidizes H_2O to create hydroxyl radicals (OH), the electron can reduce the amount of molecular oxygen to the superoxide radicals that leads to the mineralization of dyes. The photocatalytic mechanism shown in Fig. 2 is described as follows.

6 Challenges and Future Perspectives

Even though nanoparticles have demonstrated a promising results in eliminating dyes from wastewater, a number of issues still need to be resolved for their safe and efficient use. The ecotoxicity of nanomaterials in aquatic environments is one important concern. The long-term effects on the environment and possible threats to aquatic ecosystems must be assessed before the use of nanomaterials. Furthermore, use of nanoparticles on laboratory scale differs greatly from real-world industrial

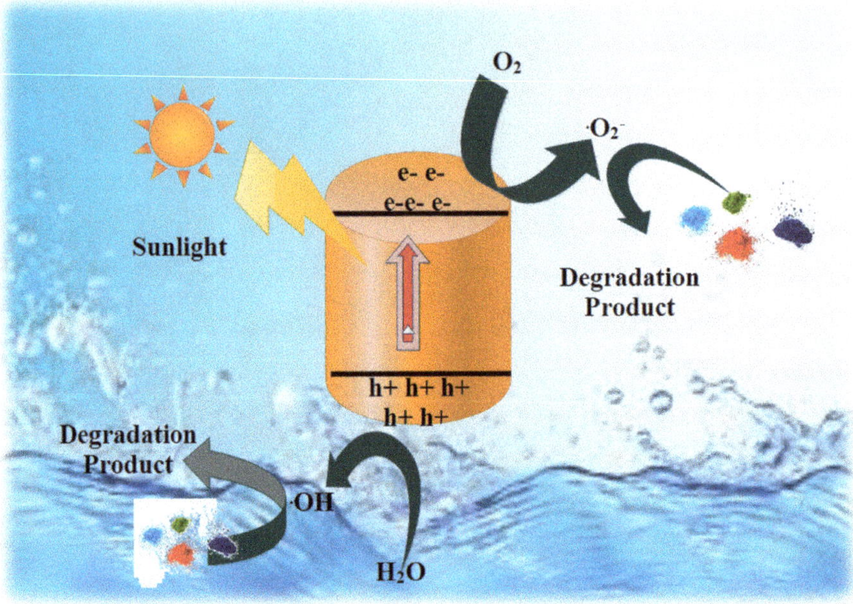

Fig. 2 Photocatalytic mechanism of degradation of toxic dyes from wastewater

treatment conditions, therefore, further studies are needed to evaluate the effectiveness of nanomaterials in actual industrial dye effluents. Regeneration and reuse of nanomaterials are additional factors that are necessary to make the process sustainable and economical. To regenerate these materials without losing their effectiveness is another great deal for a good catalysts. Environmental cleanup operations could be greatly advanced by the development of sustainable and environmentally friendly nanomaterials in the future. The need for materials that can efficiently break down pollutants while causing the least amount of ecological impact is rising as environmental issues throughout the world become more pressing. Green chemistry techniques, which emphasize the synthesis of nanomaterials utilizing non-toxic solvents, energy-efficient procedures, and renewable resources, are becoming more prevalent. These techniques provide the long-term safety of nanoparticles and suppressing the environmental impact. Moreover, a crucial aspect of the sustainable development is the formation of biodegradable nanomaterials, which decompose spontaneously without producing toxic byproducts. Another interesting approach is the integration of nanomaterials with current wastewater treatment technology. Emerging pollutants that are present in low concentrations but poses serious threats, such as industrial chemicals, medicines, and pesticides, are frequently difficult for conventional methods to remove. Treatment systems can remove pollutants more effectively by combining nanomaterials with advanced oxidation processes (AOPs), adsorption, and membrane technologies. Continuous innovation will be needed to address issues with cost effectiveness, environmental safety, and large-scale production. To ensure the safe use and disposal of nanomaterials, future studies should also concentrate on their toxicity and long-term environmental fate. Nanomaterials have the potential to completely transform environmental remediation with further development, leading to cleaner water supplies and a healthier environment.

7 Conclusion

Nanomaterials have shown a promising and transformative approach for the removal of a wide range of dyes from wastewater. The exceptional properties of nanomaterials facilitate the efficient adsorption and degradation of dyes. Ongoing research work on nanomaterials for their scalable, cost-effective, and sustainable nanomaterial-based technologies holds great potential for significantly reducing the environmental impact of pollution due to dyes. The water pollution due to dye can be overcome by using nanomaterial through the physical, chemical, and biological techniques. However by combining, physical, chemical, and biological techniques, effective results can be obtained. To conclude, nanomaterial is very appealing for wastewater treatment applications because of their excellent characteristics with distinctive morphological and structural characteristics.

References

1. Petersen L, Heynen M, Pellicciotti F (2016) Freshwater resources: past, present, future. In: International encyclopedia of geography: people, the earth, environment and technology: people, the earth, environment and technology, pp 1–1
2. Demirbas A (2008) Heavy metal adsorption onto agro-based waste materials: a review. J Hazard Mater 157(2–3):220–229
3. Arshad A, Iqbal J, Ahmad I, Israr M (2018) Graphene/Fe_3O_4 nanocomposite: interplay between photo-fenton type reaction, and carbon purity for the removal of methyl orange. Ceram Int 44(3):2643–2648
4. Ahmad A, Mohd-Setapar SH, Chuong CS, Khatoon A, Wani WA, Kumar R, Rafatullah M (2015) Recent advances in new generation dye removal technologies: novel search for approaches to reprocess wastewater. RSC Adv 5(39):30801–30818
5. Rajkumar D, Kim JG (2006) Oxidation of various reactive dyes with in situ electro-generated active chlorine for textile dyeing industry wastewater treatment. J Hazard Mater 136(2):203–212
6. Anastopoulos I, Kyzas GZ (2014) Agricultural peels for dye adsorption: a review of recent literature. J Mol Liq 1(200):381–389
7. Ahmad N, Sultana S, Faisal SM, Ahmed A, Sabir S, Khan MZ (2019) Zinc oxide-decorated polypyrrole/chitosan bionanocomposites with enhanced photocatalytic, antibacterial and anticancer performance. RSC Adv 9(70):41135–41150
8. Yagub MT, Sen TK, Afroze S, Ang HM (2014) Dye and its removal from aqueous solution by adsorption: a review. Adv Colloid Interf Sci 1(209):172–184
9. Wong YC, Szeto YS, Cheung W, McKay G (2004) Adsorption of acid dyes on chitosan—equilibrium isotherm analyses. Process Biochem 39(6):695–704

10. Malik PK, Sanyal SK (2004) Kinetics of decolourisation of azo dyes in wastewater by UV/ H_2O_2 process. Sep Purif Technol 36(3):167–175
11. Gupta VK (2009) Application of low-cost adsorbents for dye removal—a review. J Environ Manag 90(8):2313–2342
12. García-Montaño J, Domènech X, García-Hortal JA, Torrades F, Peral J (2008) The testing of several biological and chemical coupled treatments for Cibacron red FN-R azo dye removal. J Hazard Mater 154(1–3):484–490
13. Koupaie EH, Moghaddam MA, Hashemi SH (2011) Post-treatment of anaerobically degraded azo dye Acid Red 18 using aerobic moving bed biofilm process: enhanced removal of aromatic amines. J Hazard Mater 195:147–154
14. Körbahti BK, Artut K, Geçgel C, Özer A (2011) Electrochemical decolorization of textile dyes and removal of metal ions from textile dye and metal ion binary mixtures. Chem Eng J 173(3):677–688
15. Kant R (2012) Adsorption of dye eosin from an aqueous solution on two different samples of activated carbon by static batch method. J Water Resour Prot 4(2):93–98
16. Priya ES, Selvan PS (2017) Water hyacinth (*Eichhornia crassipes*)—an efficient and economic adsorbent for textile effluent treatment—a review. Arab J Chem 1(10):S3548–S3558
17. Gorio GC, Badot PM (2008) Application of chitosan, a natural minopolysaccharide, for dye removal from aqueous solutions by adsorption processes using batch studies: a review of recent literature. Prog Polym Sci 33:399–447
18. Robinson T, Chandran B, Nigam P (2002) Removal of dyes from a synthetic textile dye effluent by biosorption on apple pomace and wheat straw. Water Res 36(11):2824–2830
19. Annadurai G, Juang RS, Lee DJ (2002) Use of cellulose-based wastes for adsorption of dyes from aqueous solutions. J Hazard Mater 92(3):263–274
20. Foo KY, Hameed BH (2009) Value-added utilization of oil palm ash: a superior recycling of the industrial agricultural waste. J Hazard Mater 172(2–3):523–531
21. Hameed BH (2009) Spent tea leaves: a new non-conventional and low-cost adsorbent for removal of basic dye from aqueous solutions. J Hazard Mater 161(2–3):753–759
22. Baek MH, Ijagbemi CO, Se-Jin O, Kim DS (2010) Removal of malachite green from aqueous solution using degreased coffee bean. J Hazard Mater 176(1–3):820–828
23. Zawani Z (2009) Equilibrium, kinetics and thermodynamic studies: adsorption of Remazol Black 5 on the palm kernel shell activated carbon. Eng J Sci Res 37(1):67–76
24. Bello OS, Ahmad MA (2012) Coconut (cocos nucifera) shell based activated carbon for the removal of malachite green dye from aqueous solutions. Sep Sci Technol 47(6):903–912
25. Janoš P, Buchtová H, Rýznarová M (2003) Sorption of dyes from aqueous solutions onto fly ash. Water Res 37(20):4938–4944
26. Ruan W, Hu J, Qi J, Hou Y, Zhou C, Wei X (2019) Removal of dyes from wastewater by nanomaterials: a review. Adv Mater Lett 10(1):9–20
27. Mohmood I, Lopes CB, Lopes I, Ahmad I, Duarte AC, Pereira E (2013) Nanoscale materials and their use in water contaminants removal—a review. Environ Sci Pollut Res 20:1239–1260
28. Xu P, Zeng GM, Huang DL, Feng CL, Hu S, Zhao MH, Lai C, Wei Z, Huang C, Xie GX, Liu ZF (2012) Use of iron oxide nanomaterials in wastewater treatment: a review. Sci Total Environ 1424:1–0
29. Wang S, Sun H, Ang HM, Tadé MO (2013) Adsorptive remediation of environmental pollutants using novel graphene-based nanomaterials. Chem Eng J 15(226):336–347
30. Fan M, Li T, Hu J, Cao R, Wu Q, Wei X, Li L, Shi X, Ruan W (2016) Synthesis and characterization of reduced graphene oxide-supported nanoscale zero-valent iron (nZVI/rGO) composites used for Pb (II) removal. Materials 9(8):687
31. Cao R, Fan M, Hu J, Ruan W, Xiong K, Wei X (2017) Optimizing low-concentration mercury removal from aqueous solutions by reduced graphene oxide-supported Fe_3O_4 composites with the aid of an artificial neural network and genetic algorithm. Materials 10(11):1279
32. Zhuang Y, Ahn S, Luthy RG (2010) Debromination of polybrominated diphenyl ethers by nanoscale zerovalent iron: pathways, kinetics, and reactivity. Environ Sci Technol 44(21):8236–8242

33. Purkait MK, DasGupta S, De S (2005) Adsorption of eosin dye on activated carbon and its surfactant based desorption. J Environ Manag 76(2):135–142
34. Rafatullah M, Sulaiman O, Hashim R, Ahmad A (2010) Adsorption of methylene blue on low-cost adsorbents: a review. J Hazard Mater 177(1–3):70–80
35. Bell J, Plumb JJ, Buckley CA, Stuckey DC (2000) Treatment and decolorization of dyes in an anaerobic baffled reactor. J Environ Eng 126(11):1026–1032
36. Saratale RG, Saratale GD, Chang JS, Govindwar SP (2011) Bacterial decolorization and degradation of azo dyes: a review. J Taiwan Inst Chem Eng 42(1):138–157
37. Solís M, Solís A, Pérez HI, Manjarrez N, Flores M (2012) Microbial decolouration of azo dyes: a review. Process Biochem 47(12):1723–1748
38. Wang H, Yuan X, Zeng G, Leng L, Peng X, Liao K, Peng L, Xiao Z (2014) Removal of malachite green dye from wastewater by different organic acid-modified natural adsorbent: kinetics, equilibriums, mechanisms, practical application, and disposal of dye-loaded adsorbent. Environ Sci Pollut Res 21:11552–11564
39. Rani M, Shanker U (2023) Removal of organic dyes by functionalized nanomaterials. In: Handbook of green and sustainable nanotechnology: fundamentals, developments and applications. Springer, Cham, pp 1267–1298
40. Routoula E, Patwardhan SV (2020) Degradation of anthraquinone dyes from effluents: a review focusing on enzymatic dye degradation with industrial potential. Environ Sci Technol 54(2):647–664
41. Gürses A, Açıkyıldız M, Güneş K, Gürses MS, Gürses A, Açıkyıldız M, Güneş K, Gürses MS (2016) Dyes and pigments: their structure and properties. Dyes Pigments:13–29
42. Adenan NH, Lim YY, Ting AS (2022) Removal of triphenylmethane dyes by Streptomyces bacillaris: a study on decolorization, enzymatic reactions and toxicity of treated dye solutions. J Environ Manag 15(318):115520
43. Ghoreishi SM, Haghighi R (2003) Chemical catalytic reaction and biological oxidation for treatment of non-biodegradable textile effluent. Chem Eng J 95(1–3):163–169
44. Rajabi M, Mahanpoor K, Moradi O (2017) Removal of dye molecules from aqueous solution by carbon nanotubes and carbon nanotube functional groups: critical review. RSC Adv 7(74):47083–47090
45. Li H, Chen X, Shen D, Wu F, Pleixats R, Pan J (2021) Functionalized silica nanoparticles: classification, synthetic approaches and recent advances in adsorption applications. Nanoscale 13(38):15998–16016
46. Mohan TS, Soundharya R, Mathew BB, Anantharaju KS (2024) Photocatalytic nanomaterials and their application in tannery wastewater treatment. In: Development in wastewater treatment research and processes. Elsevier, pp 147–163
47. Bhatt P, Pandey SC, Joshi S, Chaudhary P, Pathak VM, Huang Y, Wu X, Zhou Z, Chen S (2022) Nanobioremediation: a sustainable approach for the removal of toxic pollutants from the environment. J Hazard Mater 5(427):128033
48. Smith SC, Rodrigues DF (2015) Carbon-based nanomaterials for removal of chemical and biological contaminants from water: a review of mechanisms and applications. Carbon 1(91):122–143
49. Madima N, Mishra SB, Inamuddin I, Mishra AK (2020) Carbon-based nanomaterials for remediation of organic and inorganic pollutants from wastewater. A Rev Environ Chem Lett 18:1169–1191
50. Gupta VK, Kumar R, Nayak A, Saleh TA, Barakat MA (2013) Adsorptive removal of dyes from aqueous solution onto carbon nanotubes: a review. Adv Colloid Interf Sci 1(193):24–34
51. Khan Y, Ahmad A, Ahmad N, Mir FR, Schories G (2020) Biogenic synthesis of a green tea stabilized PPy/SWCNT/CdS nanocomposite and its substantial applications, photocatalytic degradation and rheological behavior. Nanoscale Adv 2(4):1634–1645
52. Nizamuddin S, Siddiqui MT, Mubarak NM, Baloch HA, Abdullah EC, Mazari SA, Griffin GJ, Srinivasan MP, Tanksale A (2019) Iron oxide nanomaterials for the removal of heavy metals and dyes from wastewater. Nanoscale Mater Water Purif 1:447–472

53. Nagajyothi PC, Prabhakar Vattikuti SV, Devarayapalli KC, Yoo K, Shim J, Sreekanth TV (2020) Green synthesis: photocatalytic degradation of textile dyes using metal and metal oxide nanoparticles-latest trends and advancements. Crit Rev Environ Sci Technol 50(24):2617–2723
54. Ahmad N, Sultana S, Azam A, Sabir S, Khan MZ (2017) Novel bio-nanocomposite materials for enhanced biodegradability and photocatalytic activity. New J Chem 41(18):10198–10207
55. Ahmad N, Sultana S, Sabir S, Khan MZ (2020) Exploring the visible light driven photocatalysis by reduced graphene oxide supported Ppy/CdS nanocomposites for the degradation of organic pollutants. J Photochem Photobiol A Chem 1(386):112129
56. Umesh AS, Puttaiahgowda YM, Thottathil S (2024) Enhanced adsorption: reviewing the potential of reinforcing polymers and hydrogels with nanomaterials for methylene blue dye removal. Surf Interfaces 26:104670
57. Ahmad N, Anae J, Khan MZ, Sabir S, Yang XJ, Thakur VK, Campo P, Coulon F (2021) Visible light-conducting polymer nanocomposites as efficient photocatalysts for the treatment of organic pollutants in wastewater. J Environ Manag 1(295):113362
58. Dąbrowski A (2001) Adsorption—from theory to practice. Adv Colloid Interf Sci 93(1–3):135–224
59. Dawood S, Sen T ((2014) Review on dye removal from its aqueous solution into alternative cost effective and non-conventional adsorbents. J Chem Process Eng 1(104):1–1
60. Worch E (2012) Adsorption technology in water treatment. de Gruyter, Berlin
61. Tan IA, Ahmad AL, Hameed BH (2008) Adsorption of basic dye on high-surface-area activated carbon prepared from coconut husk: equilibrium, kinetic and thermodynamic studies. J Hazard Mater 154(1–3):337–346
62. Hameed BH, Ahmad AL, Latiff KN (2007) Adsorption of basic dye (methylene blue) onto activated carbon prepared from rattan sawdust. Dyes Pigments 75(1):143–149
63. Weng CH, Lin YT, Yeh CL, Sharma YC (2010) Magnetic Fe_3O_4 nanoparticles for adsorptive removal of acid dye (new coccine) from aqueous solutions. Water Sci Technol 62(4):844–851
64. Gao C, Zhang W, Li H, Lang L, Xu Z (2008) Controllable fabrication of mesoporous MgO with various morphologies and their absorption performance for toxic pollutants in water. Cryst Growth Des 8(10):3785–3790
65. Lee CK, Liu SS, Juang LC, Wang CC, Lyu MD, Hung SH (2007) Application of titanate nanotubes for dyes adsorptive removal from aqueous solution. J Hazard Mater 148(3):756–760
66. Salehi R, Arami M, Mahmoodi NM, Bahrami H, Khorramfar S (2010) Novel biocompatible composite (chitosan–zinc oxide nanoparticle): preparation, characterization and dye adsorption properties. Colloids Surf B: Biointerfaces 80(1):86–93
67. Satapanajaru T, Chompuchan C, Suntornchot P, Pengthamkeerati P (2011) Enhancing decolorization of Reactive Black 5 and Reactive Red 198 during nano zerovalent iron treatment. Desalination 266(1–3):218–230
68. Ruan W, Shi X, Hu J, Hou Y, Fan M, Cao R, Wei X (2017) Modeling of malachite green removal from aqueous solutions by nanoscale zerovalent zinc using artificial neural network. Appl Sci 8(1):3
69. Muruganandham M, Swaminathan M (2006) Advanced oxidative decolourisation of Reactive Yellow 14 azo dye by UV/TiO_2, UV/H_2O_2, UV/H_2O_2/Fe^{2+} processes—a comparative study. Sep Purif Technol 48(3):297–303
70. Ahmed AS, Ahamad T, Ahmad N, Khan MZ (2019) Removal enhancement of acid navy blue dye by GO-TiO_2 nanocomposites synthesized using sonication method. Mater Chem Phys 1(238):121906
71. Ahmad N, Sultana S, Kumar G, Zuhaib M, Sabir S, Khan MZ (2019) Polyaniline based hybrid bionanocomposites with enhanced visible light photocatalytic activity and antifungal activity. J Environ Chem Eng 7(1):102804

CMC-Fe_3O_4: Environment-Friendly Polymer Nanocomposite for the Wastewater Treatment

Jyotsna Thakur and Maheshwari Zirpe

Abstract Worldwide, industrialization led to the severe pollution causing long-term damage to the natural resources. Recent focus of water remediation technology is to develop materials which have high capacity to remove contaminants, low cost, easy retrieval and environmental compatibility. In this regard, this study aimed to prepare a nanocomposite of non-toxic and biodegradable polymer such as carboxymethyl cellulose (CMC) and magnetic Fe_3O_4 nanoparticles. The study also investigates the efficiency CMC-Fe_3O_4 synergies for the removal of toxic heavy metal like arsenic in water. CMC-Fe_3O_4 nanocomposite was prepared by dispersing as-prepared Fe_3O_4 nanoparticles in the polymer matrix in 2:1 ratio via ultrasonication. As prepared material was characterized by using X-ray diffractometer, field emission-scanning electron microscopy, energy dispersive spectroscopy, high-resolution transmission electron microscopy and Fourier transform infrared spectroscopy that confirms the formation of nanocomposite. The nanocomposite demonstrated high adsorption capacity of 287.19 $mg.g^{-1}$ towards 300 mg L^{-1} As (III) at optimum experimental conditions such as 30 min of equilibration using 0.025 g of nanoadsorbent at neutral pH. The adsorption process was well described by Langmuir isotherm model, best fitted with pseudo-second-order kinetics. Biocompatibility and easy magnetic separation of nanoadsorbent ensures the environmental safety.

Keywords Carboxymethyl cellulose-Fe_3O_4 · Nanocomposite · Wastewater treatment · Arsenic · Adsorption

J. Thakur (✉)
Changu Kana Thakur Arts, Commerce and Science College, New Panvel (w), Maharashtra, India
e-mail: thakurjyotsna@gmail.com

M. Zirpe
Ramsheth Thakur College of Commerce and Science, Kharghar, Navi Mumbai, Maharashtra, India

S. A. Bhawani and M. Jawaid (eds.), *Nanomaterials for Separation of Hazardous Contaminants from Wastewater*, Smart Nanomaterials Technology,
https://doi.org/10.1007/978-981-95-6079-0_11

1 Introduction

Scarcity of the clean water is one of the greatest challenges faced by the world today. Unprecedented growth in population, urbanization and industrial revolution is contributing to the alteration in physical, chemical and biological characteristics of water that has destructive consequences on life at large. Industries have become main source of various pollutants such as heavy metals, dyes, persistent organic pollutants, pharmaceuticals, agrochemicals, microplastics, etc. [1]. Electroplating, painting, mining, tanneries, paper industries, photographic industries, batteries, printing industries directly or indirectly discharge heavy metals such as Cr, Hg, Pb, Cd, As, Cu, Zn, Ni, etc. in water bodies [2]. In the past decade, heavy metals in water became a matter of great concern due to their toxic, non-biodegradable nature and tendency of bioaccumulation. Therefore, removal of heavy metals before being released in water resources has become essential.

Among the various heavy metals, arsenic is highly toxic metal and considered as a significant pollutant in drinking water (maximum prescribed limit by WHO is 0.05 mg L^{-1}). In environment, it is available in the form of As (III) and As (V). As (III) is considered the most toxic species; however, As (V) can easily get reduced to As (III). It is reported that it is naturally present in the groundwater of many countries [3, 4]. The poisonous properties of arsenic and its deleterious effect on human health are well studied in various countries such as India, Bangladesh, Myanmar, Canada, China, US, Thailand, etc. [5, 6].

Long-term exposure to arsenic through drinking water and food can lead to cancer and skin lesions. It may cause cardiovascular diseases and diabetes also. Exposure during pregnancy and early childhood may result in impaired cognitive development and a higher risk of death in young adults. The most crucial step is to prevent its exposure by restricting its release in environment and ensuring access to a safe and reliable water supply [7].

Number of strategies have been studied and employed for the removal of heavy metals from wastewater. Electrochemical precipitation [8], membrane separation [9], ion-exchange [10] and reverse osmosis [11] are few to name. Though these techniques are quite effective to remove heavy metals from wastewater, they have either high operational cost or require large setups or create large amount of sludge [12, 13]. The adsorption is most commonly used, economical, effective and simplest approach for removing and recycling of the water pollutants from wastewater [14, 15]. Therefore, over time, research interest in adsorbents that meet quality and environmental standards has grown remarkably. Developing adsorbents with high adsorption capacities and enhancing their properties through surface modifications has emerged as a significant research area.

1.1 Polymer Nanocomposites in Wastewater Treatment

In recent years, there has been increased utilization of nanomaterials in the environmental remediation. Nanoparticles that have high surface area, fascinating surface morphology, tunable properties, high mechanical strength can be conveniently used as nanoadsorbents, nanosensors, nanocatalysts, active membranes, etc. However, there are many challenges in using bare nanoparticles in water treatment, mainly difficulty in handling, clogging, pressure drop and recovery due to their extremely reduced size; their active surfaces result into aggregation and thus loss of activity [16]. All these limitations force to develop hybrid materials which retain the inherent properties of the nanoparticles and provide improved properties for water decontamination. In this regard, the fabrication of nanocomposite by embedding nanoparticles into larger molecules such as polymers provides an opportunity to enhance the properties such as hydrophobicity/hydrophilicity, dispersibility, porosity, colloidal stability, mechanical strength, etc. [17–19]. Nanocomposites having magnetic nature attracted the attention of the researchers to develop magnetically retrievable, environment-friendly and low-cost strategy for water treatment [20]. Therefore, the combination of Fe-based nanoparticles and any suitable polymer is the most viable option as an adsorbent. Various natural and synthetic polymers with conducting and biodegradable properties are widely used for fabrication of adsorbents. Multiple functional groups and cross-linking in polymeric material play an important role in fabrication of nanocomposites and also contribute to their adsorptive behaviour. Additionally, it facilitates ease in material handling.

1.2 CMC-Fe_3O_4 Nanocomposite as an Adsorbent

Carboxymethyl cellulose (CMC) is a derivative of cellulose biopolymer. It is water soluble, low-cost, biodegradable, non-toxic, environmentally compatible. It has strong hydrophilicity and stable internal network structure [21, 22]. CMC with functional groups such as hydroxyl and sodium carboxymethyl groups (-CH_2COONa) has potentially high adsorption capacity as it forms cross-linking with polyvalent ions [23, 24]. However, its degradability in aqueous solution limits its recyclability. Impregnation of magnetic Fe_3O_4 nanoparticles improves the recyclability and adsorption capacity of the CMC-based adsorbents. Along with the magnetic nature, Fe_3O_4 demonstrates outstanding magnetic and electronic properties and is sufficiently biocompatible. [25, 26]. Layek et al. (2013) and Li et al. (2017) reported high adsorption capacity for heavy metal ions and improved mechanical and surface energy of CMC-based nanocomposites [27, 28].

This chapter focusses on the synthesis of CMC/Fe_3O_4 nanocomposite by facile two-step method, discusses structural properties and investigates the adsorption performance with respect to adsorption capacity, adsorption rate, selectivity and reusability. The results were used to propose different isotherm, kinetic and thermodynamic models to understand adsorption process.

2 Synthetic Method of CMC-Fe_3O_4 Nanocomposite

CMC-Fe_3O_4 nanocomposite was prepared in two steps.

Step I: Fe_3O_4 nanoparticles were prepared by simple co-precipitation method by mixing aqueous solutions of $FeCl_3.6H_2O$ and $FeCl_2.4H_2O$ in 2:1 ratio. Mixture was stirred vigorously at 80 °C for 1 h. 25% NH_4OH is added dropwise to the mixture till it becomes alkaline. Black coloured precipitate formed is then washed several times to remove excess of base and dried at 80 °C for 6 h.

Step II: Certain amount of Fe_3O_4 nanoparticles were dispersed into aqueous solution of Na-CMC and sonicated for 30 min. Mixture is well stirred at 80 °C for 3 h. Magnetically separated solid then washed with number of times with distilled water followed by ethanol. Nanocomposite formed is then dried at 60 °C for 12 h and then ground into fine powder [25].

3 Characterization of CMC-Fe_3O_4 Nanocomposite

Characterization is one of the key aspects of in-depth analysis of nanocomposite. Surface and structural analysis determines size, shape, morphology, crystallinity, functional groups, surface area, porosity, surface charge of the nanocomposite which aids to understand the adsorption behaviour and the adsorption mechanism.

Crystal structure, chemical composition, lattice constants and particle size were determined by recording X-ray diffraction (XRD) spectra of CMC, as-prepared Fe_3O_4 and CMC-Fe_3O_4 nanocomposite (Fig. 1). The broad peaks at about 22° and characterizes amorphous nature of CMC (Fig. 1a). Sharp peaks of Fe_3O_4 in Fig. 1b indicate the formation of cubic phase of Fe_3O_4 which is in agreement with JCPDS file 79–0418. The average particle size of Fe_3O_4 nanoparticles was calculated using Scherrer formula which is found to be 45.79 nm. XRD spectra in Fig. 1c show the distribution of all the peaks of Fe_3O_4 nanoparticles in the CMC matrix. These results were reported in our previous research on CMC-Fe_3O_4 nanocomposite [25].

FTIR spectra (Fig. 2) reveal the information of functional groups of CMC and existence of Fe_3O_4 nanoparticles in nanocomposite. It also helps to understand adsorption mechanism. The band at 562 cm^{-1} corresponds to the Fe-O stretching vibrations in the spinel structure of magnetite. The distinctive broadband at 3342 cm^{-1} ascertained the O-H stretching vibration, while the peaks at 2927 cm^{-1}, 1392 cm^{-1} indicate the C-H stretching vibrations and bending vibration of the alkyl group, respectively. The bending vibration of -OH was observed at 1613 cm^{-1}. The strong band at 1047 cm^{-1} indicates C-O-C stretching that characterizes CMC matrix [25].

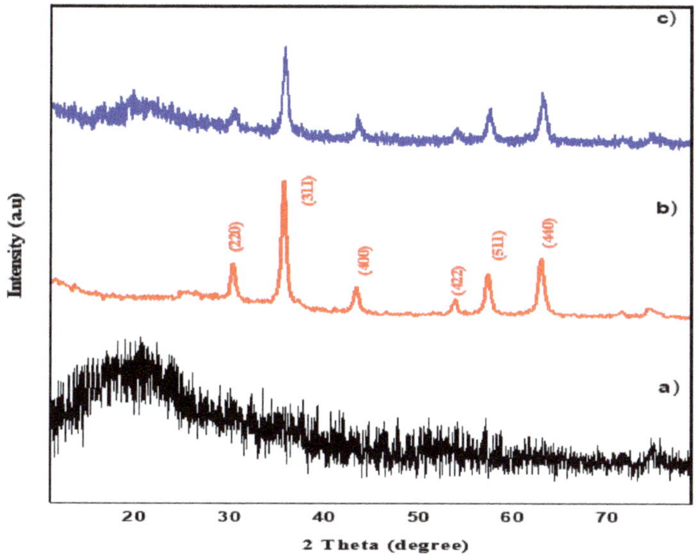

Fig. 1 XRD spectra of **a** CMC; **b** Fe_3O_4 nanoparticles; **c** CMC-Fe_3O_4 nanocomposite

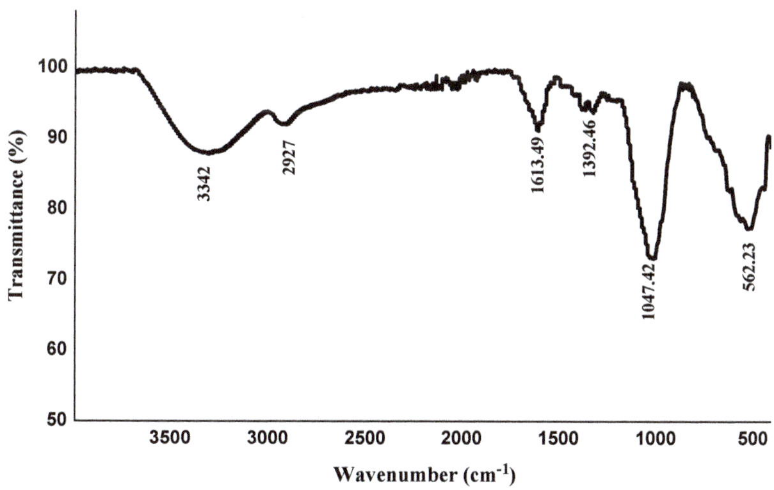

Fig. 2 FTIR spectra of CMC-Fe_3O_4

Surface morphology of CMC-Fe_3O_4 was studied before and after adsorption using FE-SEM. FE-SEM image (Fig. 3a) before adsorption shows the granular structures having size in the range of 10–20 nm. FE-SEM image (Fig. 3b) clearly indicates adsorption of As (III) on its surface. After adsorption, the surface of the adsorbent was found to be covered with adsorbate and consequently, the granular structure was disappeared.

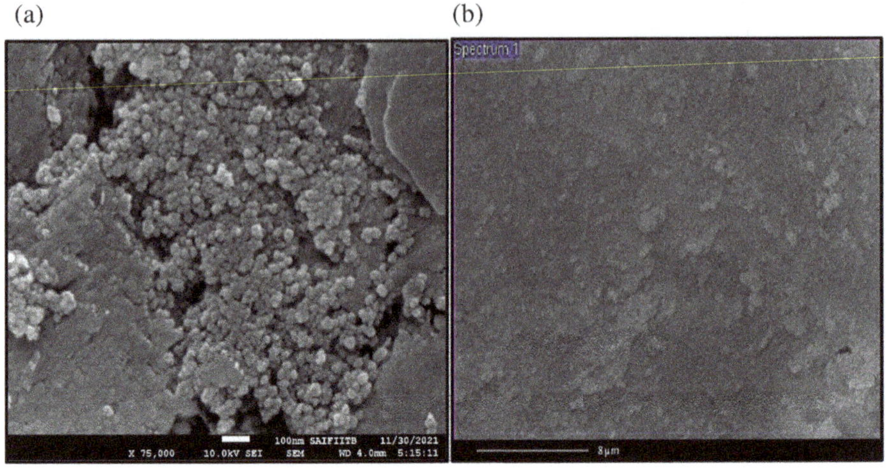

Fig. 3 FE-SEM images of CMC- Fe_3O_4 nanocomposite **a** before adsorption; **b** after adsorption of As (III) ions

EDS analysis of nanocomposite before and after adsorption study also supports the above-mentioned observation. The additional peak for As (III) in the post-adsorption spectrum (Fig. 4b) indicates successful adsorption of arsenic onto the nanocomposite. However, in Fig. 4a absence of arsenic peak suggests no detectable As (III) in the nanocomposite before adsorption that confirms effective adsorption.

The specific surface area of an adsorbent is very crucial to describe adsorption behaviour. High surface area is indicative of large number of active sites that are available for adsorption. Brunauer–Emmett–Teller (BET) using nitrogen absorption and desorption isotherm method was used to determine specific surface area of nanocomposite. The nanocomposite was found to have significantly high surface area which is 67.18 m^2g^{-1} which is reported in our previous study [25].

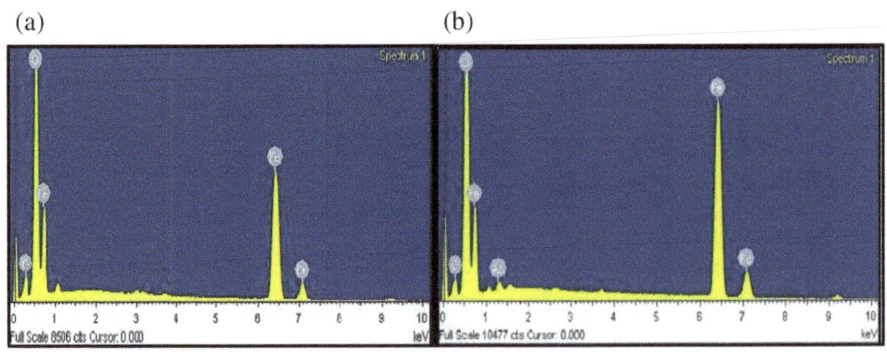

Fig. 4 EDS analysis of CMC-Fe_3O_4 nanocomposite **a** before adsorption and **b** after adsorption of As (II) ions

The pH point of zero charge (pH_{pzc}) depicts the surface charge on the nanocomposite. It is a pH value at which the material shows zero charge in aqueous medium which is measured by pH drift method [29]. pH_{pzc} value observed for CMC-Fe_3O_4 nanocomposite was 5.3 which indicates that above pH 5.3 the nanocomposite has negative surface charge.

4 Adsorption Behaviour of CMC-Fe_3O_4 Nanocomposite

As reported in our previous study, CMC-Fe_3O_4 nanocomposite has already exhibited high adsorption capacity towards heavy metals such as Cd (II) and Hg (II) [25]. The current study is focussed on the removal of another highly toxic heavy metal, As (III) from water. Adsorption of As (III) on nanocomposite was investigated by commonly used batch equilibration method. Batches of 25 mL of aqueous solution of As (III) were equilibrated with the nanoadsorbent. As CMC-Fe_3O_4 nanocomposite is magnetic in nature, it was easily separated using external magnet. Residual concentration of As (III) in effluent was measured by using ICP-AES. Percentage adsorption and adsorption capacity (Q_e) were calculated using Eqs. (1) and (2).

$$\%\ \text{adsorption} = \frac{C_0 - C_e}{C_0} \times 100 \tag{1}$$

$$Q_e = (C_o - C_e) \times \frac{V}{W} \tag{2}$$

The adsorption process is optimized to obtain maximum adsorption capacity.

5 Optimization of Adsorption Parameters

Adsorption performance of CMC-Fe_3O_4 nanoadsorbent was evaluated at different conditions to set optimum experimental condition for study.

5.1 Equilibration Time

The equilibration time for the adsorption As (III) on CMC-Fe_3O_4 nanoadsorbent was fixed by equilibrating 25 mL of 300 mgL^{-1} of As (III) solution with 0.025 g of nanoadsorbent at room temperature (RT). Equilibration time was varied from 5 to 60 min, and % adsorption is calculated for each measurement. As calculated from Fig. 5, the maximum adsorption capacity of nanoadsorbent was found to be 287.19

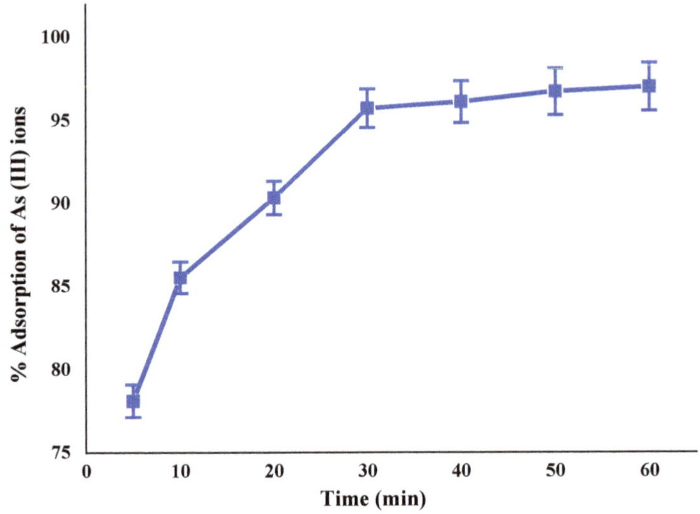

Fig. 5 Optimization of equilibration time for adsorption of As (III) ions on CMC-Fe_3O_4 nanoadsorbent

mgg^{-1} in 30 min. Thereafter, adsorption became very slow, and saturation of active sites may be the reason.

5.2 Initial Concentration

Adsorption performance of any material is based on its response to the highest concentration of target pollutants in the solution. Maximum uptake from highly concentrated solution indicates availability of large number of active sites and high surface area. In this study, 0.025 g of adsorbent was equilibrated with the concentration ranges from 100 to 700 mgL^{-1}. As seen in Fig. 6, the % adsorption decreased with the increase in concentration it may resulted due to the saturation of active sites. However, 92.73% decontamination was observed at concentration 300 mgL^{-1} indicating significantly high adsorption capacity of CMC-Fe_3O_4 towards As (III).

5.3 Amount of Adsorbent

CMC-Fe_3O_4 nanoadsorbent loading was optimized by varying amount of adsorbent (0.005–0.05 g) and equilibrating with 300 mgL^{-1} As (III) solution for 30 min. As per trend seen in Fig. 7, the adsorbent dose of 0.025 g was sufficient to remove maximum amount of arsenic from the solution. Further increase in adsorbent dose

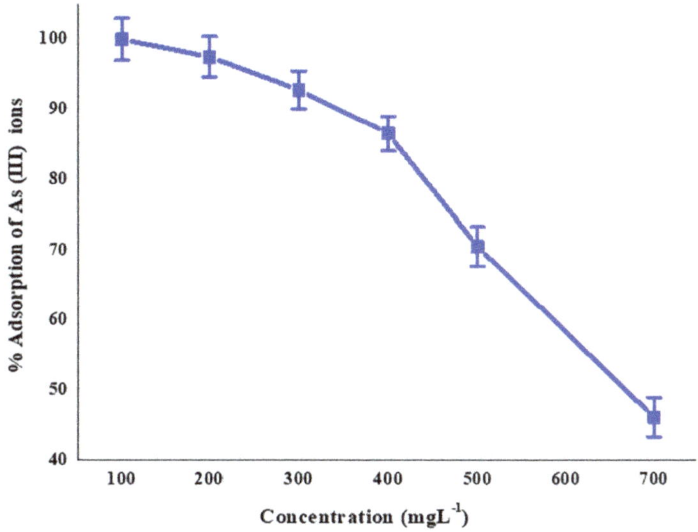

Fig. 6 Optimization of maximum concentration for adsorption of As (III) ions on CMC-Fe_3O_4 nanoadsorbent

showed minimal enhancement in adsorption which may ascribed to mass transfer limitation.

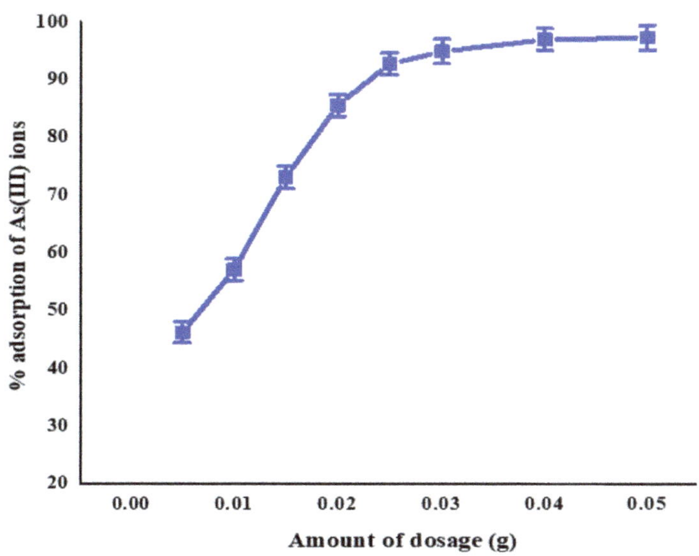

Fig. 7 Optimization of amount of CMC-Fe_3O_4 nanoadsorbent for adsorption of As (III) ions

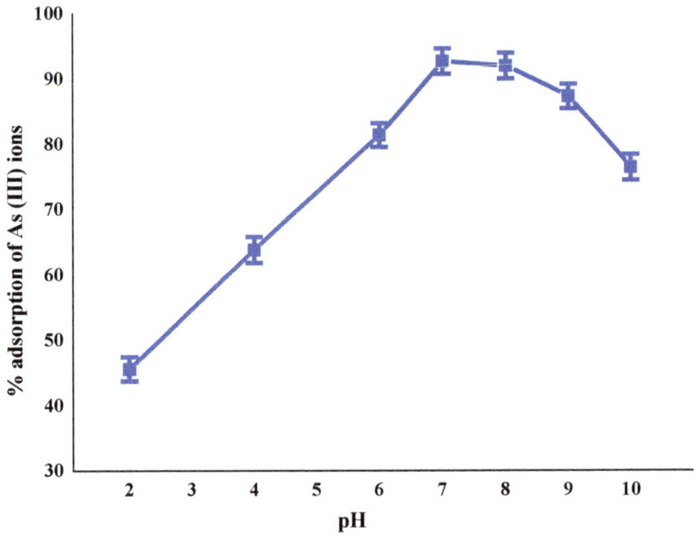

Fig. 8 Effect of pH on adsorption of As (III) ions over CMC-Fe_3O_4 nanoadsorbent

5.4 pH

As the adsorption is a surface phenomenon, surface charge plays very important role in adsorption behaviour of the adsorbent. Therefore, pH is the crucial factor to understand adsorption mechanism. Therefore, the effect of pH on adsorption process was investigated, and the results were presented in Fig. 8. It was observed that maximum adsorption of As (III) was observed at neutral pH in between 7–8.

According to pH_{pzc} value, it is expected that As (III) shows more affinity to CMC-Fe_3O_4 above pH 5.3 due to electrostatic attraction which is observed in Fig. 8. As pH shifts to acidic, H_3AsO_3 species become dominant, potentially limiting the adsorption of As (III) ions. This might be due to protonation, which could create an electrostatic repulsion force against positively charged surface of CMC-Fe_3O_4. However, in basic condition $H_2AsO_3^-$, AsO^{3-}, and AsO_3^{3-} species predominate causing electrostatic repulsion between adsorbent surface and negatively charged species [6].

5.5 Adsorption Selectivity

The effect of co-existing ion on adsorption of As (III) ions on nanoadsorbent describes the selectivity of adsorbent towards target metal ion. Water soluble other cations may compete with target metal ion for active sites. Na (I), K (I), Mg (II), Ca (II) and Al (III) were studied against As (III) and the results were presented in Fig. 9. Mg (II)

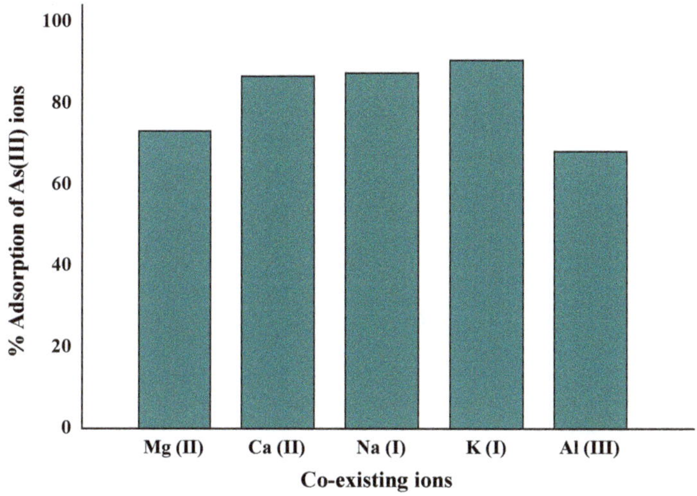

Fig. 9 Effect of co-existing ions on adsorption of As (III) ions on CMC-Fe_3O_4 nanoadsorbent

and Al (III) exhibited little competition with As (III) that may be because of their ionic size.

6 Modelling of Experimental Data

6.1 Isotherm Modelling

The experimental data was fitted in Langmuir and Freundlich isotherm models to understand the equilibrium behaviour [30, 31]. Langmuir isotherm model suggests monolayer adsorption of adsorbate on active adsorption sites of adsorbent while Freundlich isotherm model suggests sorption on the heterogeneous surface. The maximum adsorption capacity was calculated from the adsorption isotherm plot (Fig. 10a and b). The adsorption constants and variables calculated from isotherm plots are presented in Table 1.

According to Table 1, the regression co-efficient, $R^2 = 0.9992$ for Langmuir isotherm model (Fig. 10a) suggests that the adsorption data is best fitted with the Langmuir isotherm model. The maximum adsorption capacity 303.03 mgg^{-1} is also in agreement with the experimental value 287.19 mgg^{-1}. Nonlinear relationship in Freundlich model ($R^2 = 0.7084$) unable to explain adsorption (Fig. 10b).

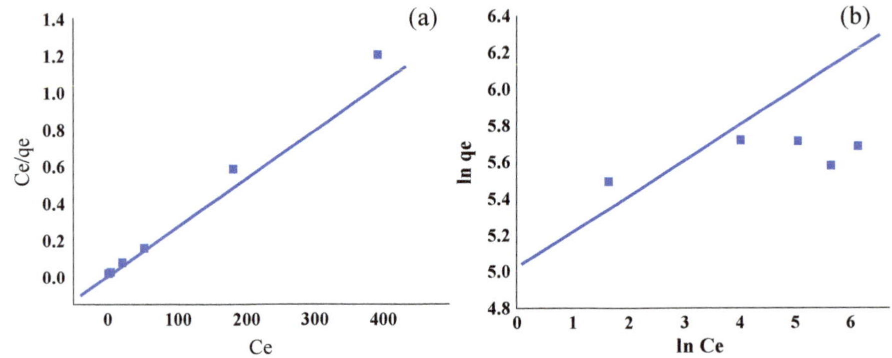

Fig. 10 **a** Langmuir Isotherm model**b** Freundlich Isotherm model for adsorption of As (III) ions on CMC-Fe_3O_4 Nanoadsorbent

Table 1 Regression co-efficient and isotherm parameters

Langmuir isotherm parameters				Freundlich isotherm parameters			Expt
Q_{max} mgg^{-1}	K_L Lmg^{-1}	R_L	R^2	K_F mgg^{-1}	n	R^2	Qe mgg^{-1}
303.03	4.7	0.00071	0.9992	121.51	5.136	0.7084	287.19

6.2 Kinetic Modelling

Kinetic modelling evaluates the rate of adsorption and the interactions between adsorbate and active sites of adsorbent that influence the adsorption mechanism [32, 33]. In this study, the experimental data was investigated with pseudo-first-order and pseudo-second order kinetic models as represented in Fig. 11a,b.

Kinetic parameters calculated from the plots in Fig. 11 are summarized in Table 2. It is clearly appeared that pseudo-second-order plot exhibits linearity with $R^2 = 0.9998$ and dominated over pseudo-first-order kinetics. The equilibrium adsorption capacity (qe) 294.12 mgg^{-1} is in accordance with the experimental value. The result suggests that the adsorption process is chemisorption driven.

6.3 Thermodynamic Study

The adsorption rate is highly influenced by temperature. Generally, the adsorption rate is increased with temperature. The Gibbs free energy change ΔG^o is the fundamental criterion of the spontaneity of a reaction, whereas ΔH^o expresses standard enthalpy change. Both the parameters were determined from Van't Hoff plot in Fig. 12. The Van't Hoff thermodynamic parameters are summarized in Table 3.

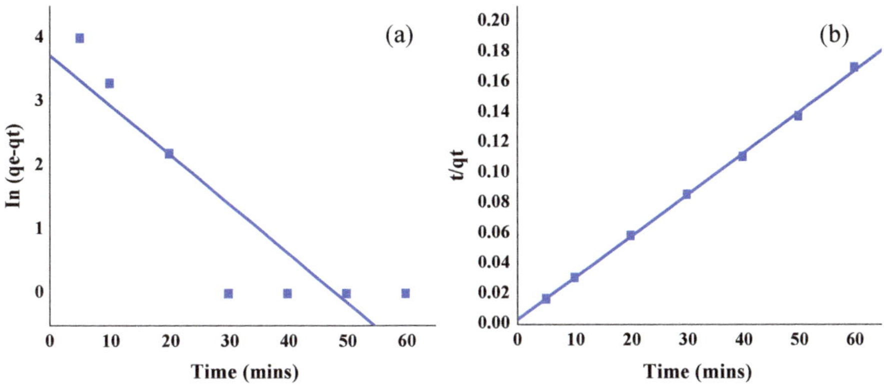

Fig. 11 **a** Pseudo-first order **b** pseudo-second-order kinetic models for adsorption As (III) ions on CMC-Fe_3O_4 nanoadsorbent

Table 2 Kinetic parameters

Pseudo 1st order			Pseudo 2nd order			Expt
K_1 min^{-1}	Q_e mg g^{-1}	R^2	K_2 min^{-1}	qe mg g^{-1}	R^2	qe mg g^{-1}
0.0804	50.75	0.7984	0.0019	294.12	0.9998	287.19

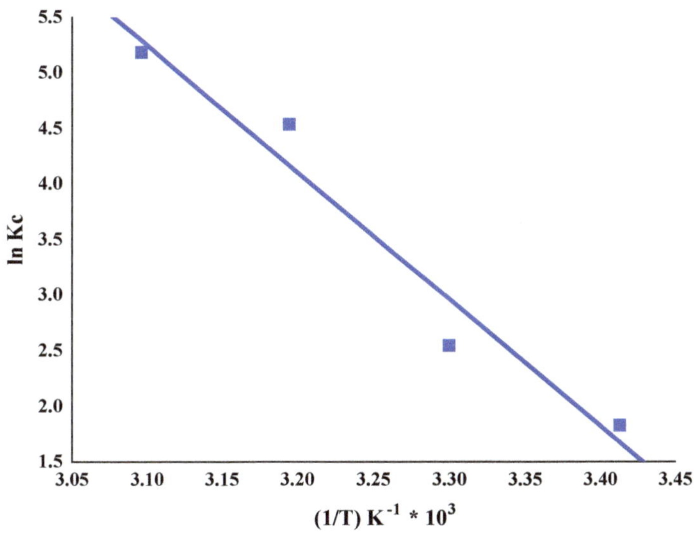

Fig. 12 Van't Hoff plot for adsorption of As (III) on CMC-Fe_3O_4

Table 3 Thermodynamic parameters

ΔG^o ($kJmol^{-1}$)				ΔH°	ΔS°
298 K	303 K	313 K	323 K	$kJmol^{-1}$	J mol^{-1} K^{-1}
− 4.45	− 6.41	− 11.8	− 13.92	94.86	337.79

The negative value of ΔG^o confirms the spontaneous nature and feasibility of the adsorption of As (III) on CMC-Fe_3O_4 nanoadsorbent. The positive value of ΔH^o and ΔS^o indicate the endothermic nature of the adsorption process and increased randomness at the interface during interaction between As (III) ion and the nanoadsorbent [34, 35].

7 Regeneration Study

In recent years, there is a critical need for highly efficient adsorbents that can be regenerated and reused to minimize unnecessary waste generation. In this regard, regeneration/reusability study of adsorbent becomes very crucial. Figure 13 illustrates the reusability of CMC-Fe_3O_4 nanoadsorbent through an adsorption–desorption study.

Adsorption of As (III) ions at optimum condition was followed by desorption using different desorbing agents such as 0.1 M HNO_3, 0.1 M NaOH and 1% Alum. The CMC-Fe_3O_4 nanoadsorbent was dispersed in the desorption agent and magnetically stirred for 3 h at room temperature to regenerate the nanoadsorbent. The regenerated CMC-Fe_3O_4 nanoadsorbent was separated using magnets and rinsed with double distilled deionized water. The adsorption–desorption process is repeated for number of cycles. After every cycle, the residual As (III) ion concentration in the aqueous phase was determined. The results showed that 0.1 M NaOH acted as a better desorbing agent that effectively recycled the adsorbent for four successive cycles.

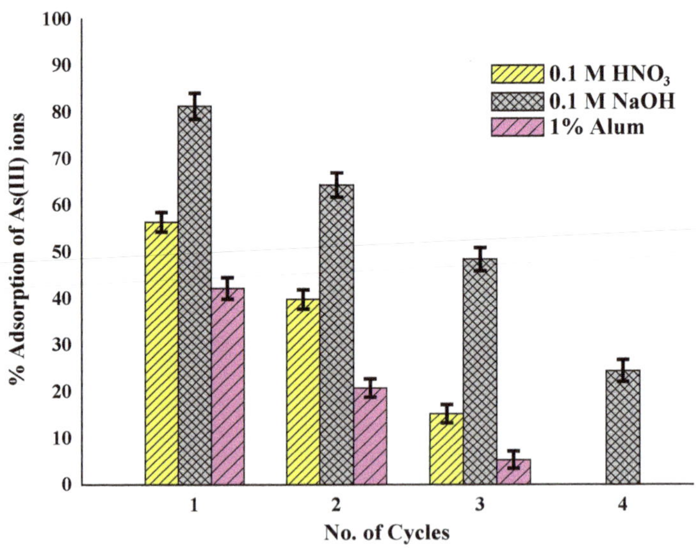

Fig. 13 Regeneration and reusability study of CMC/Fe_3O_4 nanoadsorbent for removal of As (III) ions

8 Conclusion

CMC-Fe_3O_4 nanocomposite demonstrated remarkable potential for rapid removal of As (III) ions from the aqueous solution. It was synthesized by simple co-precipitation method assisted with ultrasonication. Its structure and morphology support adsorption behaviour of the nanocomposite, and its magnetic nature facilitates its easy removal from wastewater. Nanocomposite possesses high surface area (67.18 m^2g^{-1}) that improves the efficiency of nanocomposite as adsorbent.

The maximum adsorption capacity of adsorbent was found to be 287.19 mgg^{-1} for As (III) at neutral pH for 300 mgL^{-1} As (III) solution within 30 min. The adsorption process was well described by Langmuir isotherm model and pseudo-second-order kinetic model indicating single-layer adsorption on the homogeneous surface of adsorbent and chemisorption, respectively. Thermodynamic parameters calculated from adsorption study suggest the feasible and spontaneous adsorption process. CMC-Fe_3O_4 nanocomposite thus proved to be an efficient and promising nanoadsorbent for As (III) in wastewater treatment.

References

1. Kumar M, Borah P, Devi P (2020) Priority and emerging pollutants in water. In: Inorganic pollutants in water, pp 33–49
2. Owa FW (2014) Water pollution: sources, effects, control and management. Int. Lett. Nat. Sci. 3:1–6
3. Wang S (2006) Mulligan CN occurrence of arsenic contamination in canada: sources, behavior and distribution. Sci Total Environ 366:701–721
4. Garelick H, Jones H, Dybowska A, Valsami-Jones E (2009) Arsenic pollution sources. In: Reviews of environmental contamination, pp 197, Springer New York NY. https://doi.org/10.1007/978-0-387-79284-2_2
5. Kapaj S, Peterson H, Liber K, Bhattacharya P (2006) Human health effects from chronic arsenic poisoning—A review. J Environ Sci Health A Tox Hazard Subst Environ Eng 41(10):2399–428. https://doi.org/10.1080/10934520600873571
6. Shankar S, Shanker U, Shikha (2014) Arsenic contamination of groundwater: a review of sources prevalence health risks and strategies for mitigation. Sci World J 2014:304524. https://doi.org/10.1155/2014/304524
7. WHO, Arsenic in groundwater 2022. https://www.who.int/news-room/fact-sheets/detail/arsenic
8. Bazrafshan E, Mohammadi L, Ansari-Moghaddam, A et al (2015) Heavy metals removal from aqueous environments by electrocoagulation process—A systematic review. J Environ Health Sci Engineer 13:74. https://doi.org/10.1186/s40201-015-0233-8
9. Xiang H, Min X, Tang C-J, Sillanpää M, Zhao F (2022) Recent advances in membrane filtration for heavy metal removal from wastewater: a mini review. J Water Process Eng 49:103023
10. Hafez A, El-Mariharawy S (2004) Design and performance of the two-stage/twopass RO membrane system for chromium removal from tannery wastewater. Part 3 Desalination 165:141–51
11. Zhu J, Wei S, Chen M, Gu H, Rapole SB, Pallavkar S et al (2013) Magnetic nanocomposites for environmental remediation. Adv Powder Technol 24(2):459–467
12. Bhandari H, Garg S, Gaba R (2021) Advanced nanocomposites for removal of heavy metals from wastewater. Macromol Symp 397(1):2000337. https://doi.org/10.1002/masy.202000337

13. Ray PZ, Shipley HJ (2015) Inorganic nano-adsorbents for the removal of heavy metals and arsenic: a review. RSC Adv 5:29885–29907
14. Zirpe M, Thakur J (2022) Sono-assisted synthesis of $AgFeO_2$ nanoparticles for efficient removal of Basic Green-4 dye from aqueous solution. J Nanopart Res 24:240. https://doi.org/10.1007/s11051-022-05620-2
15. Zirpe, M., Thakur, J (2023) Spindle-shaped $AgFeO_2$ nanoparticles for efficient removal of cationic and anionic dyes. J Disp Sci Technol, pp 1–12. https://doi.org/10.1080/01932691.2023.2283100
16. Daer S, Kharraz J, Giwa A, Hasan SW (2015) Recent applications of nanomaterials in water desalination: a critical review and future opportunities. Desalination 367:37–48
17. Bhaumik M, McCrindle RI, Maity A (2015) Enhanced adsorptive degradation of congo red in aqueous solutions using polyaniline/Fe0 composite nanofibers. Chem Eng J 260:716–729. https://doi.org/10.1016/j.cej.2014.09.014
18. Zirpe M, Bagla H, Thakur J (2019) Adsorptive removal of fluoride using polymer-modified ceria nanoparticles: determination of equilibrium, kinetic and thermodynamic parameters. Sep Sci Technol 55(16):2933–2947. https://doi.org/10.1080/01496395.2019.1660674
19. Zirpe M, Bagla H, Thakur J (2020) Rapid removal of $^{152+154}Eu$ (III) using polyaniline/ceria nanocomposite from low level waste. J Inorg Organomet Polym 30:5053–5062. https://doi.org/10.1007/s10904-020-01606-6
20. Mehmood A, Khan FSA, Mubarak NM et al (2021) Magnetic nanocomposites for sustainable water purification—A comprehensive review. Environ Sci Pollut Res 28:19563–19588. https://doi.org/10.1007/s11356-021-12589-3
21. Rees DA (1969) Structure, conformation, and mechanism in the formation of polysaccharide gels and networks. Adv Carbohydr Chem Biochem 24:267–332
22. Thomas M, Naikoo GA, Sheikh MUD, Bano M, Khan F (2016) Effective photocatalytic degradation of Congo red dye using alginate/carboxymethyl cellulose/TiO_2 nanocomposite hydrogel under direct sunlight irradiation. J Photochem Photobiol Chem 327:33–43. https://doi.org/10.1016/j.jphotochem.2016.05.005
23. Wu L, Lin X, Zhou X, Luo X (2016) Removal of uranium and fluorine from wastewater by double-functional microsphere adsorbent of SA/CMC loaded with calcium and aluminum. Appl Surf Sci 384:466–479. https://doi.org/10.1016/j.apsusc.2016.05.056
24. Liu Y, Wang W, Wang A (2010) Adsorption of lead ions from aqueous solution by using carboxymethyl cellulose-g-poly (acrylic acid)/attapulgite hydrogel composites. Desalination 259(1–3):258–264
25. Zirpe M, Thakur J (2023) Magnetically recoverable CMC/ Fe_3O_4 nanocomposite for efficient removal of Hg(II) and Cd(II) ions from the aqueous solution. Sep Sci Technol 58(15–16):2692–2703. https://doi.org/10.1080/01496395.2023.2225733
26. Luna-Martínez JF, Reyes-Melo E, González-González V, Guerrero-Salazar C, Sepúlveda-Guzmán A (2013) Synthesis and characterization of a magnetic hybrid material consisting of iron oxide in a carboxymethyl cellulose matrix. J Appl Polym Sci 127(3):2325–2331
27. Layek RK, Kundu A, Nandi AK (2013) High-performance nanocomposites of sodium carboxymethyl cellulose and graphene oxide. Macromol Mater Eng 298(11):1166–1175
28. Li WY, Zuo P, Xu DD et al (2017) Tunable adsorption properties of bentonite/carboxymethyl cellulose-g-poly(2-[dimethylamino] ethyl methacrylate) composites toward anionic dyes. Chem Eng Res des. 124:260–70
29. Stumm W (1992) Chemistry of the solid-water interface: processes at the mineral-water and particle-water interface in natural systems. Wiley-Interscience, New York
30. Langmuir I (1916) The constitution and fundamental properties of solids and liquid. J Am Chem Soc 38:2221–2295
31. Freundlich HMF (1906) Over the adsorption in solution. J Phy Chem 57:385–470
32. Langergen S (1989) About the theory of so-called adsorption of soluble substances. Kungl Svenska Vetensk Acad Handl, pp 1–39
33. Ho YS, McKay G (1999) Pseudo-second order model for sorption process. Process Biochem 34:451–465

34. Saha P, Chaudhari, S (2011) Insight into adsorption thermodynamics. In: Themodynamic, editor M. Tadashi INTECH open access publisher UK, pp 349–364
35. Gupta VK (1998) Equilibrium uptake, sorption dynamics, process development, and column operations for the removal of copper and nickel from aqueous solution and wastewater using activated slag, a low-cost adsorbent. Ind Eng Chem Res 37:192–202